DINOSAUR BEHAVIOR

Published by Princeton University Press
41 William Street, Princeton, New Jersey 08540
99 Banbury Road, Oxford OX2 6JX
press.princeton.edu

Copyright © 2023 UniPress Books Limited
www.unipressbooks.com

All rights reserved. No part of this book may be reproduced or transmitted in any form or by any means, electronic or mechanical, including photo-copying, recording, or by any information storage and retrieval system, without written permission from the copyright holder.

Requests for permission to reproduce material from this work should be sent to permissions@press.princeton.edu

Publisher: Nigel Browning

Commissioning editor: Kate Shanahan

Project editor: Katie Crous

Infographics: Sarah Skeate

Picture researcher: Tom Broadbent

Cover artwork by Bob Nicholls

Cover design by Wanda España

Library of Congress Control Number 2023932856

ISBN 978-0-691-24429-7

Ebook ISBN 978-0-691-25496-8

Printed in Slovakia

British Library Cataloging-in-Publication Data is available

10 9 8 7 6 5 4 3 2 1

DINOSAUR BEHAVIOR

AN ILLUSTRATED GUIDE

MICHAEL J. BENTON

**ILLUSTRATED BY
BOB NICHOLLS**

PRINCETON UNIVERSITY PRESS

PRINCETON AND OXFORD

CONTENTS

Introduction 7

1

DINOSAURS IN PERSPECTIVE 9
Why are Dinosaurs Important? 10
Geological Time 12
Ancient Geographies 16
Ancient Reptiles in the Sea and Air 21
Knowing About the Past 28
FORENSICS: Managing the Dig Site 30
Origins of Dinosaurs 32
The Three Main Dinosaur Groups 36
Birds as a Branch of Theropods 40

2

PHYSIOLOGY 45
Visualizing Dinosaurs 46
FORENSICS: Putting Flesh on the Bones 51
Modern Reptiles and Birds 52
A Special Kind of Warm-Bloodedness 54
Feathers 58
Breathing 64
Food Budgets 68
FORENSICS: Making a Dinosaur Food Web 70

3

LOCOMOTION 75
Posture and Gait 76
FORENSICS: Modeling Locomotion 78
Two Legs or Four? 80
Trackways 82
FORENSICS: Tippy-Toes Take-Off 87
Calculating Running Speed 88
FORENSICS: Dinosaur Walking Kinematics 90
Giant Size and Support 92
FORENSICS: How to be a Giant 97
Origins of Flight 98

4

SENSES AND INTELLIGENCE 105
Dinosaur Brains 106
FORENSICS: Studying Ancient Brains 108
Brains of Modern Reptiles and Birds 110
Evolving Intelligence 114
Sense of Smell 120
Sight 122
Hearing 124
Spatial Orientation 126

5

FEEDING 133
Dinosaur Diets 134
Teeth and Jaws 138
FORENSICS: Finite Element Analysis of Jaws 141
Evidence of Dinosaur Diets 142
FORENSICS: Isotopes and Diet 150
Dinosaur Herbivory 152
Dinosaur Carnivory (and Other Diets) 158

6

SOCIAL BEHAVIOR 163
Communication and Interaction 164
Sexual Selection and Feather Color 166
FORENSICS: Telling the Color of Fossil Feathers 172
Courtship and Mating Behavior 174
Eggs and Parental Care 176
Growing Up 184
FORENSICS: Measuring Growth Rate 188
Communication and Living in Herds 190
FORENSICS: Sounds of the Hadrosaurs 194

7

DINOSAURS AND HUMANS 197
Extinction 198
FORENSICS: A Whimper or a Bang? 206
Survivors and the Modern World 208
Opportunities 210

BIBLIOGRAPHY 218
INDEX 220
ACKNOWLEDGMENTS 224

INTRODUCTION

T. REX ENJOYS A FEAST OF BABY TRICERATOPS
The 5-ton predator should have been able to kill and eat any plant-eating dinosaur of its day, but an adult *Triceratops* had a bony frill protecting its neck, and three massive horns on its face, and it almost certainly used these in defense.

T. rex crashes through the trees and leaps on a juvenile *Triceratops*. The huge predator has scaly skin and tufts of colorful feathers over its eyes; it crunches through the bones of the terrified herbivore with a force of several tons . . .

We are familiar with this kind of scene from the movies, but how much of it is guesswork? When we look at dinosaur behavior, we're examining every aspect of how these remarkable beasts lived. Do we need a time machine to be sure what they looked like and how they behaved? The answer is, not necessarily. We have fossils, and we have smart ways to compare those fossils to modern animals.

It's true that early debates about the running ability of dinosaurs were indeed based on guesswork. One professor might say *T. rex* was slow-moving, whereas another might say it could gallop at speed, like a huge ostrich. Now, we can calculate its running speed at 17 miles (27 km) per hour—really quite slow—either by measuring its stride length from fossil tracks or by using known relationships of leg muscle dimensions and maximum speeds from modern animals and estimated muscle dimensions of the dinosaur.

These accurate insights into dinosaur behavior are valuable because dinosaurs dominated the land for over 160 million years. It's important to understand how they constructed their food webs (who eats what), how they could be so huge, and how they interacted with ancestors of modern groups (such as crocodiles, lizards, birds, and mammals).

How did dinosaurs behave with each other? Did they ignore other dinosaurs, or did they form family groups, caring for their young? Did they behave more like crocodilians or like birds, their living relatives? For example, did some or all dinosaurs show sexual display behavior, where males danced and preened to show off their beautiful tails, crests, and other adornments, as many birds do?

Paleontologists name a new dinosaur species every ten days, which is the highest level of new discoveries ever in this field. But when we're not digging up dinosaur fossils we're studying them, plugging gaps in our knowledge and trying to understand how dinosaurs lived and loved, fought and fed, signaled and interacted with one another. This book presents the knowledge we've gained, through science and discovery, about these fascinating giants that once roamed the earth.

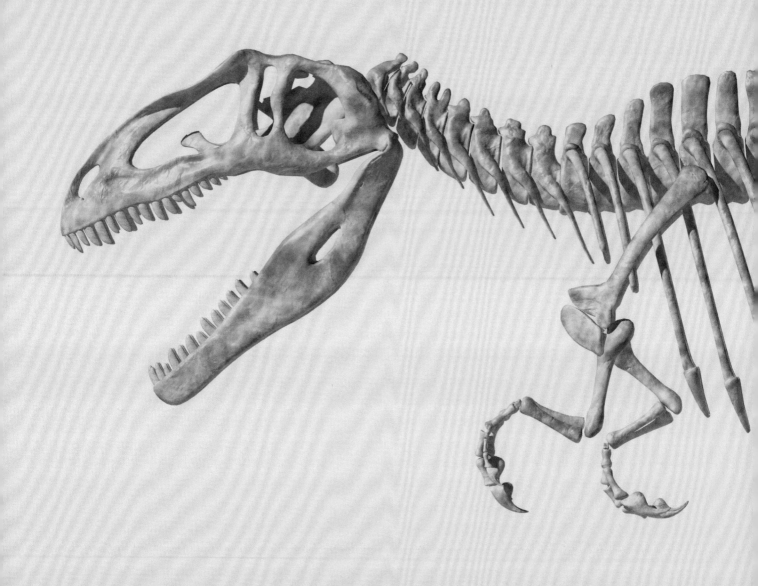

1
DINOSAURS IN PERSPECTIVE

WHY ARE DINOSAURS IMPORTANT?

Most people love dinosaurs, children in particular, but we should all care about them for two scientific reasons: they are an integral part of our history, and they stretch the limits of the possible.

There are more books for children about dinosaurs than most other subjects. Dinosaurs are popular because they were big, and some were very, very, very big. They are exciting and resembled dragons. They had amazing horns and spikes, and some had scary, huge teeth. Unlike dragons, however, they were real.

But more than being popular, dinosaurs are actually important for science. First, they are an important part of the history of Earth and life. Many big questions in science are about origins—the origin of the universe, Earth, life, and humans. Finding out about these beginnings are core components of the sciences of geology (the study of Earth) and paleontology (the study of ancient life). We need to know the details, and the best way to achieve that is to study fossils and rocks, and to understand geological time (see page 12).

A second reason we care about dinosaurs is that they are a major step on the way to the modern world. People are concerned about threats to Earth and life due to human activity. As the numbers of people on Earth rise higher and higher, as we drive

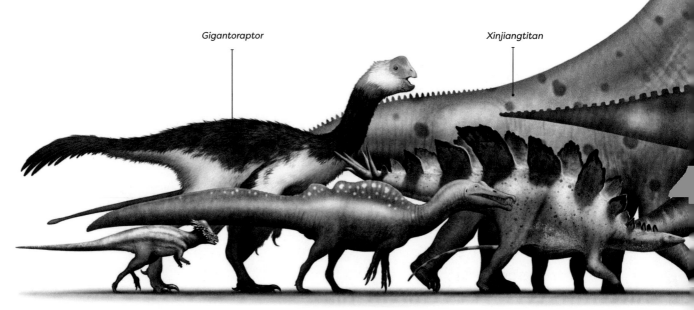

Gigantoraptor

Xinjiangtitan

Pachycephalosaurus

Ichthyovenator

Stegosaurus

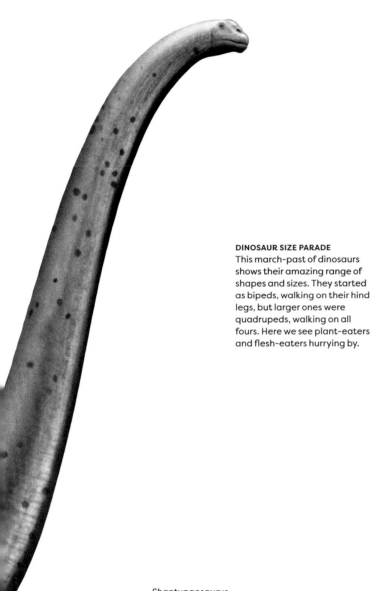

more cars and use more resources for our lives, we are damaging nature. Nature consists of all the 10 million or so species on Earth, plus their natural habitats. These species have a history, and we need to understand their history to be able to understand the threats to their survival. We will look at the origin of dinosaurs (see page 32) and find that they arose at a time when all of life was modernizing and becoming more energetic than before.

The third reason we care about dinosaurs is that they are awesome. There are no land animals today that are so huge: an elephant may weigh 5 tons, but some of the long-necked dinosaurs such as *Diplodocus* and *Brontosaurus* weighed 50 tons. How could such a huge animal function? Among animals in the days of the dinosaurs were the pterosaurs (see page 24), and some of them were much larger than any modern bird yet they could fly. So, dinosaurs stretch the limits of biomechanics (see page 88): we have to understand the math and physics of how they could be so huge and how their flesh and bone functioned.

DINOSAUR SIZE PARADE
This march-past of dinosaurs shows their amazing range of shapes and sizes. They started as bipeds, walking on their hind legs, but larger ones were quadrupeds, walking on all fours. Here we see plant-eaters and flesh-eaters hurrying by.

Shantungosaurus

Sapeornis

Bajadasaurus   *Euoplocephalus*   *Plateosaurus*   *Torosaurus*

GEOLOGICAL TIME

Earth is ancient, but geologists can work out the order of rocks as well as establishing exact dates and ages—we just have to get used to talking about *millions* of years.

Geologists talk about "millions" and "hundreds of millions" of years. Earth is 4,567 million years old, or about 4 billion years old. Dinosaurs lived from 240 to 66 million years ago. How do we know these ages?

At first, geologists could work out the *order* of the rocks, that is, what is older and what is younger. This can be done by looking at the pile of rocks in the side of a quarry or in a gorge or sea cliff. Usually, the oldest rocks are at the bottom, and younger and younger rocks are above; we see the rocks simply in the order in which they were laid down. Here, we are mainly considering sedimentary rocks, such as mudstone, sandstone, and limestone. These were once sediments, such as sand, silt, or mud, that were deposited in ancient seas, lakes, rivers, and deserts.

The order is one thing, but what about the age? The first clues come from fossils. Over two hundred years ago, geologists began to try to understand the rocks they cared about, such as those containing valuable minerals like coal and iron. Industrialists wanted to be sure they would find coal or iron when they dug a pit, and geologists found smart ways to tell them where to dig. They identified, for example, that in Europe and North America most of the coal occurs at only one part of the geological column, or the pile of rocks, in an interval they called the "Carboniferous," which means "coal bearing." The Carboniferous was identified by unique fossil plants and other fossils.

By about 1850, nearly all the main divisions of time had been named: the geological periods such as Carboniferous, Triassic, Jurassic, and Cretaceous. They were named after some key feature of rocks (such as coal) or after a region where they occur: Jurassic, for example, was named for the Jura mountains at the border of France and Switzerland. The Triassic, Jurassic, and Cretaceous together are called the Mesozoic, meaning "middle life."

LAYERS OF HISTORY
The rock succession through the Grand Canyon in Arizona documents over a billion years of Earth's history, from Precambrian at the base, through Palaeozoic, to Mesozoic. Here, the rocks at the top are Permian in age.

THE MARCH OF TIME

This chart of geological time back to the beginning of the dinosaurs shows the main time divisions and their ages in millions of years.

Dinosaurs lived through the Triassic, Jurassic, and Cretaceous, and famously died out at the very end of the Cretaceous.

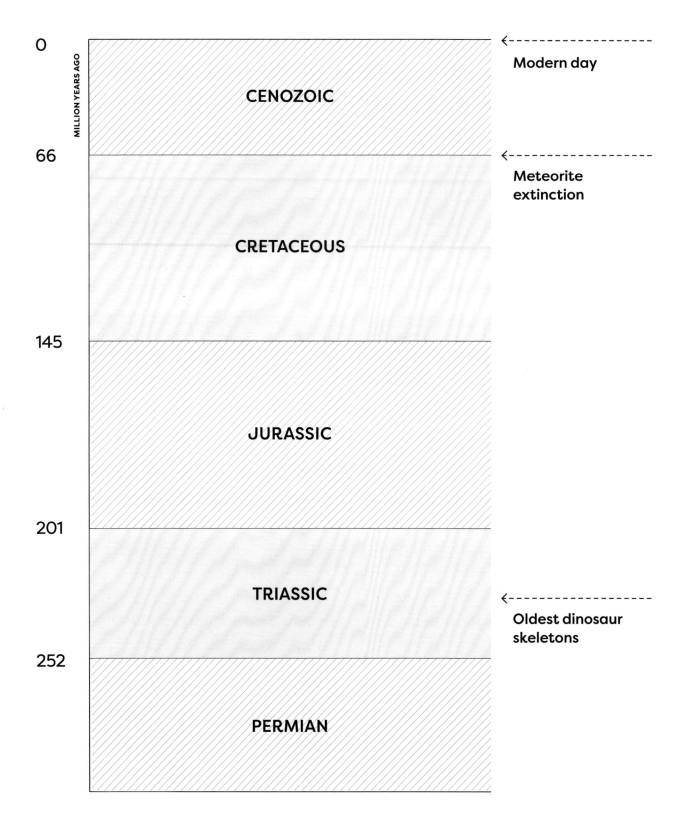

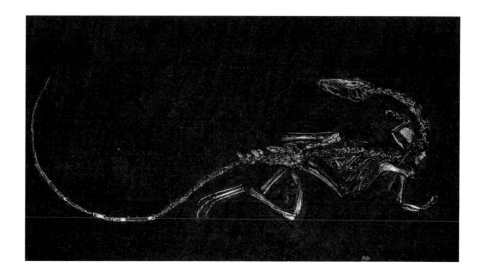

THE SKELETON OF AN EARLY DINOSAUR
A skeleton of a *Coelophysis*, from the Late Triassic of New Mexico, showing its head and long neck bent back, and the long whip-like tail. It ran on its hind legs and grabbed prey with its strong hands.

ESTIMATING THE AGE OF EARTH

We now know that the Earth is billions of years old, but this was not always the case. In the late 1800s, scientists tried all kinds of ways to estimate the age of rocks. For example, some tried to estimate how long it would take for all the hundreds and thousands of miles of sedimentary rocks to be deposited. Others assumed that the oceans had once been filled with fresh water, and so they estimated the age of the oceans by their degree of saltiness today and how long it would take all those salts to be weathered off rocks on land and washed into the oceans. These gave estimates of hundreds of millions of years.

Then the physicist Lord Kelvin (1824–1907) argued that Earth had started as a ball of molten rock, and he calculated how long it would take for such a huge ball to cool down and for the crust to form. He looked at experiments with cannon balls and his final estimate was that Earth was about 20 to 40 million years old.

This was too short a period for geologists: it did not allow enough time to fit in all the geological eras and all the fossils and their evidence. Geologists needed a more reliable method to estimate the age of Earth, and that method was invented over a hundred years ago.

RADIOMETRIC DATING

In about 1910, geologists realized they could use the natural radioactivity of certain rocks to establish exact ages. Some elements such as uranium are naturally radioactive, meaning they are unstable and change to more stable forms, in this case the metal lead. The rate of change (usually called radioactive decay) is known from experiments and may last for thousands or millions of years. The point when half the uranium in a sample has decayed to lead marks its half-life, which, for different kinds of uranium, is 4,500 million, 700 million, or 25,000 years. Geologists can measure the proportions of uranium and lead in a rock and then calculate the age depending on how far it has progressed along the decay curve. The ages are checked and rechecked all the time to make sure they are accurate.

ANCIENT GEOGRAPHIES

The world was very different in the past, thanks to continental drift and changing sea levels. In the age of dinosaurs, most continents were joined together.

Over spans of millions and billions of years, it is perhaps no surprise to discover that the geography of Earth has changed a great deal. For a start, sea levels have varied significantly, and we are concerned today that they are rising due to the warming of the climate. The predicted sea-level rise will flood major cities such as New York and London.

However, sea levels today are relatively low because huge volumes of water are locked up in the ice at the North and South Poles. In the past there were times when there was no ice, sometimes called "hothouse worlds." This is true of most of the Mesozoic, when the dinosaurs existed. In fact, in the Late Cretaceous (100 to 66 million years ago) sea levels were as much as 650 ft (200 m) higher than today, and this reduced the sizes of the continents by flooding over all the coastal plains. In addition, Africa and North America were each divided in two by huge seaways. In North America, the Western Interior Seaway was up to 600 miles (970 km) wide and linked the Caribbean to the Arctic Ocean, flooding across Texas, Wyoming, and Alberta.

There has been another huge driver of changing geography through time, or paleogeography. This is continental drift. Over two hundred years ago, geographers noticed that the west coast of Africa and the east coast of South America matched quite closely. What if the South Atlantic Ocean closed up? In fact, they had spotted something, because those two coasts had once been joined and there was no South Atlantic Ocean in the Triassic or Jurassic.

A German geographer called Alfred Wegener (1880–1930) suggested in 1912 that all the southern continents—South America, Africa, Antarctica, India, and Australia—had once been fused together as the supercontinent Gondwana. He used paleontological and geological evidence, noticing how specific rocks connected across all those continents, but also shared Permian and Triassic plants and animals. In fact, there had been a single supercontinent called Pangea in the Permian and Triassic, made from Gondwana, and a northern continent made from North America, Europe, and Asia.

At the end of the Triassic and in the Jurassic, the North Atlantic began to unzip,

LIFE IN THE WESTERN INTERIOR SEAWAY of North America, 70 million years ago. Some giant, predatory mosasaurs hunt the swimming *Triceratops* that are hurrying across a coastal bay to seek some plant food on the other side.

DINOSAURS IN PERSPECTIVE

CRETACEOUS

The continents continued to move apart in the Cretaceous (145–66 million years ago), and the plants and animals on each continent began to become increasingly different. The South Atlantic opened, and the great southern continents, including Africa, India, Australia, and Antarctica, began to move toward their present positions.

JURASSIC

In the Jurassic (201–145 million years ago), some oceans had begun to open up, splitting Pangaea with the great Tethys Ocean around the equator. Also, the North Atlantic had begun to open, with North America and Europe drifting apart by about half an inch (1 cm) per year. Dinosaurs and other animals could still wander between Africa and North America but their land routes were becoming restricted.

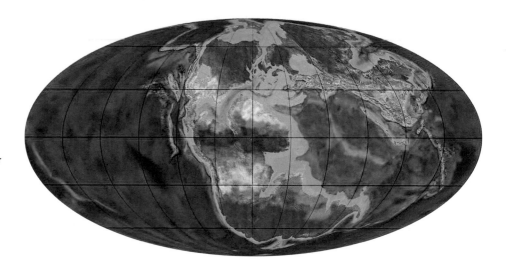

TRIASSIC

In the Triassic (252–201 million years ago), all continents were fused as the supercontinent Pangaea. Also, there were no polar ice caps, and temperatures worldwide were generally warm in summer. Early dinosaurs and other land animals and plants could move across all the lands.

followed by the South Atlantic in the Cretaceous. Antarctica moved south, Australia moved east, and India began a long journey as an island, finally joining the rest of Asia 40 million years ago. These ideas were controversial until the 1960s, when the plate tectonic engine was discovered. This explains how new crust is formed by mid-ocean ridges on the seabed, pushing the continents apart by about half an inch (1 cm) per year.

THE WESTERN INTERIOR SEAWAY

One of the most remarkable discoveries of geology in North America was that the entire continent was split by a great seaway some 80 million years ago. In the Late Cretaceous, global sea levels were 650 ft (200 m) higher than today, and the sea spilled over continental margins, making the coastline of North America move quite a distance inland. The Caribbean was much larger, and Washington and New York would have been completely submerged.

But, running from Mexico and Texas in the south, through Oklahoma, Colorado, Kansas, Wyoming, Montana, Alberta, and the Northwest Territories was a great seaway, 600 miles (970 km) wide, over 2,000 miles (3,200 km) long, and up to 2,500 ft (760 m) deep in the center. This was the Western Interior Seaway and it lasted for 50 million years, through the last days of the dinosaurs, their extinction, and into the early parts of the age of mammals in the Paleocene.

How did this affect dinosaurs? It meant that dinosaurs could no longer trek long distances east and west. Up to the time of rising sea levels, a dinosaur could wander from California to North Carolina, but now the western and eastern dinosaur populations were isolated. Instead of terrestrial deposits full of dinosaurs and plants, we find marine limestones in those states where the sea passed full of fossil fish and marine reptiles. Populations of *T. rex* and *Triceratops* on the western shores could only gaze across hundreds of miles of sea to their cousins in the east.

PLATE TECTONICS

New Earth's crust is formed at the mid ocean ridges, making the ocean floor about half an inch (1 cm) wider each year. So, for example: the Atlantic widens each year; in the north, North America and Europe move apart; in the south, South America and Africa are moving away from each other.

ANCIENT REPTILES IN THE SEA AND AIR

At the same time as the dinosaurs, ichthyosaurs and plesiosaurs hunted in the oceans, and leathery-winged pterosaurs flew in the skies.

Dinosaurs were land dwellers. They all could probably swim of course, just like most modern land animals, but their swimming would have been simply to cross rivers or lakes to find new feeding grounds or to escape from a predator. Perhaps the large predator *Spinosaurus* was an active swimmer, hunting fish in the Early Cretaceous rivers and lakes of North Africa. Also, in giving rise to birds, some of the birdlike dinosaurs could also probably fly. But there were other groups of reptiles that swam and flew in the Mesozoic.

The swimmers were mainly ichthyosaurs and plesiosaurs. The ichthyosaurs came on the scene at the beginning of the Triassic, and quickly established themselves as a new kind of predator in the oceans. Ichthyosaurs were shaped like sharks or dolphins, pointed at the front and with a round, smooth body, resembling something like a torpedo. This shape allowed them to swim efficiently by beating their deep-finned tail from side to side. They had large paddles at the front but these were for steering.

In the Triassic, ichthyosaurs ranged in size from a small salmon to a blue whale, 85 ft (26 m) long, and they must have followed a wide range of lifestyles. The smaller ichthyosaurs presumably hunted small fish, and larger ichthyosaurs would have chased larger fish and indeed other marine reptiles. Ichthyosaurs had long, slender snouts armed with rows of long, pointy teeth. It's probable that they swam into a shoal of fish and snapped their jaws fast, hoping to spear some fish on their teeth. As they snapped their jaws shut, water was expelled to the sides, and any unfortunate fish were caught in a cage of long teeth. Some of the whale-sized ichthyosaurs may have filtered their prey in great gulps of water and food.

GIVING BIRTH
The ichthyosaur *Nannopterygius*, from Middle Jurassic Europe and Russia, in the act of giving birth. The baby is one of six to ten in her belly and it is ready to swim and hunt as soon as it is born. Note that the tail comes out first, so that as soon as the snout is free, the baby can rush to the surface and take a gulp of air.

Plesiosaurs had four paddles for swimming, and some had long necks and small skulls; others had massive skulls and short necks. The long-necked plesiosaurs hunted fish by darting their snaky necks around quickly. The large-headed pliosaurs crunched through large reptiles as their favorite food. There were many other kinds of marine reptiles, all of them air-breathers and all of them tussling over the rich resources of fish and other sea food.

WARM BLOOD AND LIVE BIRTH

Two key things about the ichthyosaurs and plesiosaurs were that they were warm-blooded and they gave birth to live young. At one time, scientists assumed that all the Mesozoic reptiles, including ichthyosaurs, plesiosaurs, and dinosaurs, were cold-blooded like modern crocodiles and lizards (see page 52). However, study of their bone structure shows that the marine reptiles, like modern sea turtles, actually regulated their body temperature to keep it fairly constant, and indeed they were able to generate heat inside their bodies. This was necessary so they could keep up active swimming even in very cold sea waters.

These marine reptiles also gave birth to live young at sea, just as modern dolphins and whales do. This was not certain for a long time because it was reasonable to assume they crept onto land, just as sea turtles do today, to lay their eggs in the sandy beaches. In fact, wonderful specimens of ichthyosaurs had been found over 150 years ago with baby ichthyosaurs inside their rib cages. At first, paleontologists debated whether these were actually unborn babies or whether the large ichthyosaurs had been cannibals, feeding on their babies!

Cannibalism proved a popular idea, but the evidence said otherwise. The fact is that the skeletons of babies inside adult ichthyosaurs are complete, with no sign of being broken, which would surely not be the case if they had been bitten, chomped, and swallowed. Further, there can be large numbers of babies inside, sometimes four or six, even as many as ten or eleven. When there are multiple babies inside, they tend to be arranged neatly, running parallel to the adult's backbone, which is evidence they are inside the mother's body and inside her uterus.

The baby ichthyosaurs were born tail-first, not headfirst as with humans and most mammals. Whales show the same flip of birthing position, and the reason is obvious. When a baby is born into air, it comes headfirst so it can get its first gasp of air before it has completely left the mother's body.

Underwater, an air-breathing mammal or reptile would suffocate before it could get to the surface if it were born headfirst. So, out they wiggled tail-first, then at the very last minute, as they popped completely out, they would race to the surface for a gasp of air.

The mothers with unborn babies are much more common than might be expected by chance. Some even seem to show the mother in the process of giving birth, with perhaps one of the babies expelled and lying beside the mother. However, what seems to have happened is that the pregnant mothers were in danger near to the time of birth. When something disastrous happened, they died, still with the babies inside, but the gases of decomposition released as the mother's body decayed on the seabed sometimes blew one of the babies out.

COELOPHYSIS
Dinosaurs were almost certainly warm-blooded, and many of them had feathers, even if only short bristly ones for insulation. The tail of this dinosaur moved from side to side with its stride.

MESOZOIC FLYERS

The pterosaurs also lived through the Late Triassic, Jurassic, and Cretaceous, starting out quite small—about the size of a thrush or pigeon—and ending with some absolute giants, such as *Quetzalcoatlus* from the Late Cretaceous of Texas. This airplane-sized pterosaur was as tall as a giraffe when it stood upright and had a wingspan up to 36 ft (11 m).

Pterosaurs fed on fish by swooping over shallow seas, and some fed on insects. We don't know all their diets for sure, and it can be hard to determine, especially because many of them lost their teeth through evolution. At one time, this might have been thought to be good evidence that they were not predators, but of course today toothless birds such as eagles and vultures are very successful meat-eaters.

New evidence shows that pterosaurs had feathers. It was long known that pterosaurs were covered in fluff made from short whiskers, called pycnofibers. The fluff could be seen in many fossils collected in Europe in the 1800s, and paleontologists were not surprised; after all, these were presumably active animals that could fly just as well as any bird, and so they must have been warm-blooded. Flight takes a lot of energy, and pterosaurs must have eaten a lot to keep the inner furnaces going. Being insulated with fur is a good adaptation to keep the heat in, just as we see in modern birds and mammals.

PTEROSAURS AS BIG AS A PLANE
Some Late Cretaceous pterosaurs were huge, and some were really huge. Here a *Wellnhopterus*, with a wingspan of 12 ft (3.5 m), flies past its relative, the truly vast *Quetzalcoatlus*, with a wingspan of 36–39 ft (11–12 m). This giant stood as tall as a giraffe, and it seems incredible that it could have flown.

But feathers? In fact, in 2019 and 2022 two studies were published, one concerning Jurassic pterosaurs from China, the other the giant *Tupandactylus* from the Early Cretaceous of Brazil, which showed the pycnofibers were more than just simple whiskers. Some of them showed complex side branches, like those of bird feathers, and so these were interpreted as feathers.

As we will see later (page 58), many dinosaurs had feathers, too, and these new discoveries show that feathers originated perhaps 100 million years earlier, in the ancestor of birds and pterosaurs, long before the first true birds emerged.

NEXT PAGE
An *Edmontosaurus* carcass has floated out to sea in the Early Cretaceous warm seas on the coast of Brazil, and it is proving to be of great interest to flocks of birds and pterosaurs in the skies above, and plesiosaurs, hesperornithid birds, and fish underwater. One dinosaur carcass could provide tons of meat for scavengers.

DINOSAURS IN PERSPECTIVE

KNOWING ABOUT THE PAST

We learn about the past from fossils, but we must understand how they are buried and how they are collected; sometimes even soft tissues such as the guts or skin may be preserved.

The word fossil means "dug up," and at one time people used the term for anything out of the ground—rocks, arrowheads, potatoes. Now we mean specifically the remains of ancient microbes, plants, and animals. Fossils are usually the so-called hard parts such as shells or skeletons. For dinosaurs, we may find the skeleton, and the bones are usually scattered.

We can reconstruct the story behind a typical dinosaur skeleton you might see in a museum, let's say a *Tyrannosaurus rex*. The animal lived in what is now North America, and our individual might have died as an oldster, perhaps aged forty. After the animal dies, it lies beside a river, and all kinds of scavengers, such as mammals, crocodiles, and beetles, come and strip the flesh from the bones. A storm might blow up and the river level rises, eventually rolling the skeleton along, with bits of skin and flesh still trailing in the water.

Then, after a few hours, the storm stops and the river level falls, and the remaining bones fall to the riverbed at a sandy bend. By now, the head has rolled another mile downstream, and one leg has torn away, leaving just parts of the skeleton still joined together. More sand comes down the river and covers the bones. After a few years, the river has changed its course and soil forms, then more sand builds up. The thickness of sediment above the bones presses down and squeezes out water, and eventually the sand and bones turn to rock, with deposition of minerals in the tiny spaces inside the bones.

Then, 66 million years later, badland erosion in Montana cuts a small canyon across these old river sands. The great thicknesses of rock deposited in this area in the latest Cretaceous have been named the Hell Creek Formation, and there are other fossils near to the *Tyrannosaurus* skeleton—fish, insects, leaves, even a pond turtle, and an early mammal. A farmer bounces across the land in his utility vehicle and spots some white, shiny rocks in the side of the ravine. He walks over and recognizes they are bones. He knows about dinosaurs because a crew from a local university excavates on his ranch land. He calls Professor Smith, and she promises to visit the next summer with her students and excavate the skeleton.

In special cases, more than the bones may be found. Some of the badland dinosaurs show impressions of their skin in the rock. Other, smaller fossils may be preserved in shallow lakes where the gentle deposition of mud means that even some traces of the guts or feathers may be found. These are the fossils every paleontologist treasures, because they can say so much about how the dinosaurs lived.

PIECED TOGETHER
A classic museum skeleton, showing *T. rex* on the prowl. But when it was found, this skeleton was probably incomplete and lying scattered, where it had been rolled and washed by an ancient river.

FORENSICS:
MANAGING THE DIG SITE

Paleontologists don't just dig up bones as if digging up potatoes. If they rush to lift the bones, many will break and a great deal of information could be lost. Paleontologists work like archeologists, mapping and recording everything as they go, and they might work with other scientists. For example, a geologist might be on site to record all the rock details that can give indications about the ancient environment. There might also be a paleobotanist on site looking for leaves, roots, seeds, and other plant fossils.

Usually when some bones are found, it isn't immediately clear how far they go into the rock. The team may remove the overburden, which is the rock above the level with bones. It's best to work on a flat site, and the workers take great care as they come down toward the bone level, removing small amounts of rock with small drills, picks, and other hand tools.

If the bones are large, even yards long, each one has to be explored individually, removing rock above and round the sides. They are left in place so the team can photograph the whole site, sometimes using a drone if the site is large. These images can be very useful later, when the bones are back in the lab, to understand how the different blocks fit together.

The workers then dig around the larger bones so they are lying on a narrow pedestal of rock. They protect the bones with cloth or paper, and soak strips of burlap (sackcloth) in a watery plaster mixture. They then lay burlap strips, crisscrossing over the

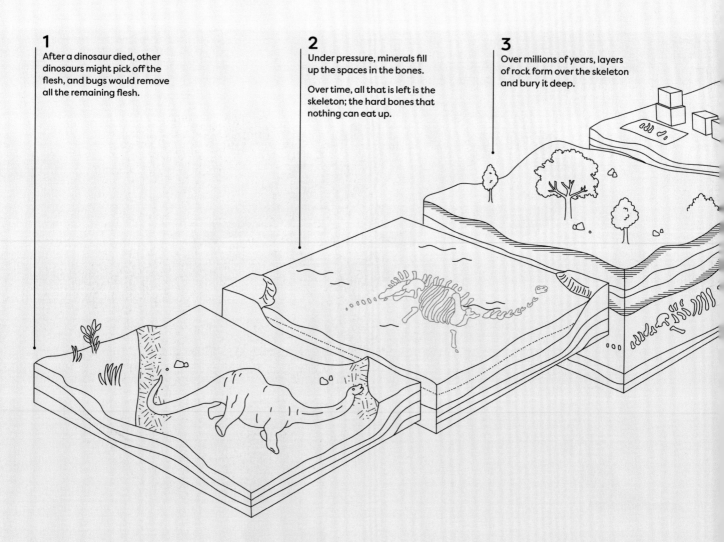

1
After a dinosaur died, other dinosaurs might pick off the flesh, and bugs would remove all the remaining flesh.

2
Under pressure, minerals fill up the spaces in the bones.

Over time, all that is left is the skeleton; the hard bones that nothing can eat up.

3
Over millions of years, layers of rock form over the skeleton and bury it deep.

bones, and let the plaster set hard. By building up many layers of plaster and burlap, the bone is parceled up, except underneath where it stands on the rock pedestal.

Then, the paleontologists carefully knock the strengthened bone package off the pedestal and parcel up the underside with more plaster and burlap, and after it sets, they can carry the bone away. If the parceled bone is very heavy, maybe weighing a ton or more, the team have to use lifting equipment.

Sometimes a large dinosaur has to be divided into many parcels, each weighing several tons, and it might take many trucks to transport them all back to the museum or university laboratory. Then, back in the lab, and using the aerial photos and maps, they put the whole thing together in order to understand how the pieces combine to form a skeleton.

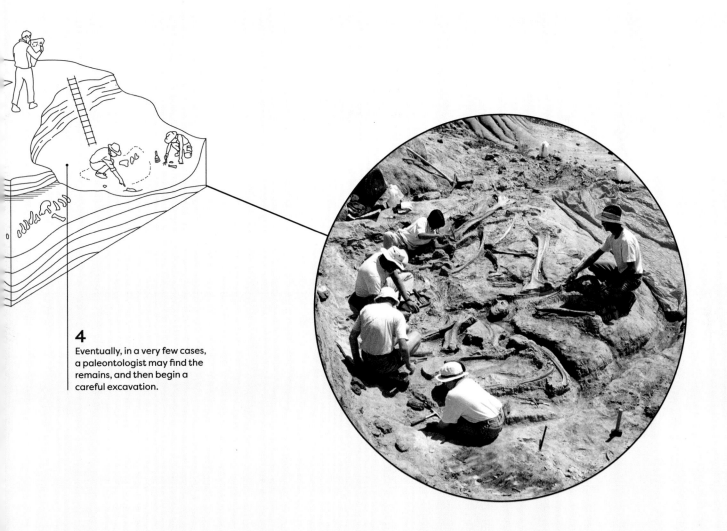

4
Eventually, in a very few cases, a paleontologist may find the remains, and then begin a careful excavation.

ORIGINS OF DINOSAURS

Dinosaurs are first found in the Triassic period, and it seems they originated in the Early Triassic following a huge mass extinction; they were warm-blooded from the start.

There was a time before the dinosaurs, in fact most of the history of Earth. The oldest fossils of dinosaurs come from the Ischigualasto Formation in northwest Argentina. In the famous Valley of the Moon, hundreds of skeletons of ancient reptiles have been found, and these include remains of a few different dinosaurs, including *Eoraptor* and *Herrerasaurus*. These are dated to 227 million years ago, based on radiometric dating (see page 15) of volcanic ash beds.

Eoraptor was a two-legged dinosaur about 3 ft (1 m) long, with a small head and long neck. Its sharp little teeth suggest it had a mixed diet of plants and flesh, and it used its powerful hands to grab leaves or small animals. *Herrerasaurus* was a much more terrifying animal, 10 ft (3 m) long, and probably a dedicated predator, shredding smaller animals on its sharp teeth. These two lived

LATE TRIASSIC, ARGENTINA
This period would have seen some of the oldest dinosaurs we know. A large *Herrerasaurus* chases a small pack of *Eoraptor*. Even these oldest dinosaurs likely had a fuzz of simple feathers, although the fossils do not show this was definitely the case.

side by side with small lizard ancestors, early mammals, maybe even some early pterosaurs, as well as large predators related to modern crocodiles.

These are the oldest skeletons of dinosaurs we know, so does this mark the origin of the group? In fact, it does not, because dinosaur-like footprints have been found in Early and Middle Triassic rocks, as old as 247 million years. Skeletons of close relatives of dinosaurs are also found in rocks of the same age, so this means there is a 20 million-year time span when we know there must have been dinosaurs on Earth—but we haven't found them yet!

It turns out that the dinosaurs originated as part of a major restructuring of ecosystems that had been triggered by a mass extinction. At the end of the Permian, 252 million years ago, there was a huge crisis caused by the largest series of volcanic eruptions ever. Only 5 percent of species survived, and these few survivors set the basis for a whole new world at the beginning of the Triassic. In the oceans, new kinds of shellfish, lobsters, and fish appeared, and they were hunted by new marine reptiles such as ichthyosaurs (see page 21).

On land, there were big changes in the survivors. The Late Permian ecosystems were

dominated by reptiles of all sizes, and they were nearly all cold-blooded, like modern reptiles, and they walked with a sprawling gait, with their limbs out at the side. At the start of the Triassic, the survivors mainly switched to a parasagittal gait, with the limbs beneath the body (see page 76). This enabled the new beasts of the Triassic to run faster, and they had become a little warm-blooded, so they had better endurance. From the start, it seems that even the earliest dinosaurs already had an insulating pelt of bristly little feathers.

MASS EXTINCTIONS AND DINOSAUR ORIGINS

It is well known that the dinosaurs were finally killed off by a mass extinction at the end of the Cretaceous (see page 198), but it is quite a new idea that their *origins* were in mass extinctions; indeed, three mass extinctions seem to have been involved!

The first was at the end of the Permian, 252 million years ago, when huge volcanic eruptions in Russia caused shocking climate changes; the air and sea became hotter, and acid rain fell from the skies. The combined effects were to kill forests on land and coral reefs at sea, and all life had to flee from the tropical belt, heading north and south to cooler climates. In the time of turmoil, as life recovered from this awful event, the first tiny dinosaurs emerged on the scene, but they were very rare.

Then, 232 million years ago, another extinction event hit, also driven by massive volcanic eruptions, this time in western Canada. Again, climates changed sharply and life was hit hard; climates became dryer and plants changed dramatically. Many plant-eaters died out and the dinosaurs increased in numbers because they could survive in the new, dry conditions on Earth.

Finally, in the third step to domination, a third mass extinction, 201 million years ago, at the end of the Triassic, killed off the large flesh-eating crocodile-like reptiles. The causes were similar, this time with volcanic eruptions in the Central Atlantic, as the North Atlantic Ocean began to open up. After this crisis, finally, the flesh-eating dinosaurs, the theropods, had their chance to become large and to prey on the big plant-eating sauropods.

So, famously, the dinosaurs were eventually wiped out by a mass extinction at the end of their reign, but their initial steps to take over the world, in the Triassic and Early Jurassic, were driven by three great mass extinctions.

ORIGIN OF FEATHERS

Birds originated 150 million years ago, in the Late Jurassic, and that was long thought to be the point at which feathers originated. Now, remarkably well preserved fossils from China show us that many dinosaurs also had feathers, and so too did their cousins, the pterosaurs. So, now we know that the origin of feathers was in fact 100 million years earlier than paleontologists had thought, at 250 million years ago, when life was recovering from the great end-Permian mass extinction.

| 252 | 201 | 145 | 66 |

MILLION YEARS AGO

| TRIASSIC | JURASSIC | CRETACEOUS |

Birds

Theropods

Sauropods

Ornithischians

Pterosaurs

Origin of feathers is 100 million years earlier than originally thought

Recovery of life from end-Permian mass extinction

DINOSAURS IN PERSPECTIVE

THE THREE MAIN DINOSAUR GROUPS

Dinosaurs fall into three main groups, all of which arose in the Triassic, and they expanded in diversity as they generally became increasingly large through the Jurassic.

The first dinosaurs, such as *Eoraptor* and *Herrerasaurus,* all looked similar, and they are distinguished as members of Dinosauria by their hind legs. As the dinosaurs stood upright, just as birds and mammals do today, their legs became modified. Compared to their sprawling ancestors, the ankle and knee joints had become simple hinges, essentially allowing movement backward and forward, but not sideways. The hip also modified to accommodate the thigh bone entering from directly below, rather from the side.

The third feature of dinosaurs and their close relatives is that the metatarsals, the long bones in the feet, were bunched and lifted from the ground. In fact, dinosaurs stood on the tips of their toes, technically called digitigrade. Birds and many mammals such as cats, dogs, and horses are also digitigrade. Humans and bears, like most reptiles, on the other hand, keep the soles of their feet, including the metatarsals, flat on the ground, technically called plantigrade.

WALKING HADROSAUR
Shantungosaurus walks on all fours but can also run on its hind legs with its hands clear of the ground. It stands on the tips of its toes and fingers: the digitigrade stance.

These features allowed the warm-blooded early dinosaurs to take long strides and chase their prey successfully. From the start, dinosaurs split into two groups, then three: the Saurischia and Ornithischia mark the fundamental split, and they are recognized by the layout of their hip bones. In saurischians, the pubis bone stuck out forward, whereas in ornithischians it swung back. Still in the Late Triassic, the Saurischia split into Theropoda (all flesh-eating dinosaurs) and Sauropodomorpha (large to very large long-necked plant-eaters).

The Ornithischia split into many familiar dinosaur groups in the Jurassic and Cretaceous, including the armored ankylosaurs, the plate-backed stegosaurs, the unarmored ornithopods, and later the horned ceratopsians and thick-headed pachycephalosaurs. The theropods also split into many groups in the Jurassic, including the birds.

WALKING TALL

Reptiles today and in the past typically have a sprawling posture (below left), where the limbs stick out to the side, and a step follows a sideways and forward movement. In dinosaurs, as in mammals, the limbs are under the body, in the erect, or parasagittal, posture (middle and right). The limb bones fit into the shoulder and hip girdles in different ways, but in both cases from below, and strides are simply back and forward (parasagittal) without a sideways swing.

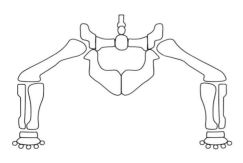
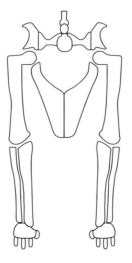
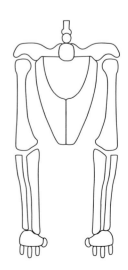

DINOSAURS IN PERSPECTIVE

EVOLUTION OF DINOSAURS

Here we chart the evolution of the dinosaurs, from the Late Triassic to their extinction at the end of the Cretaceous. The width of the spindle shapes indicates the diversity of each group through time; notice how birds (far left) nearly died out at the end of the Cretaceous, but then bounced back afterward.

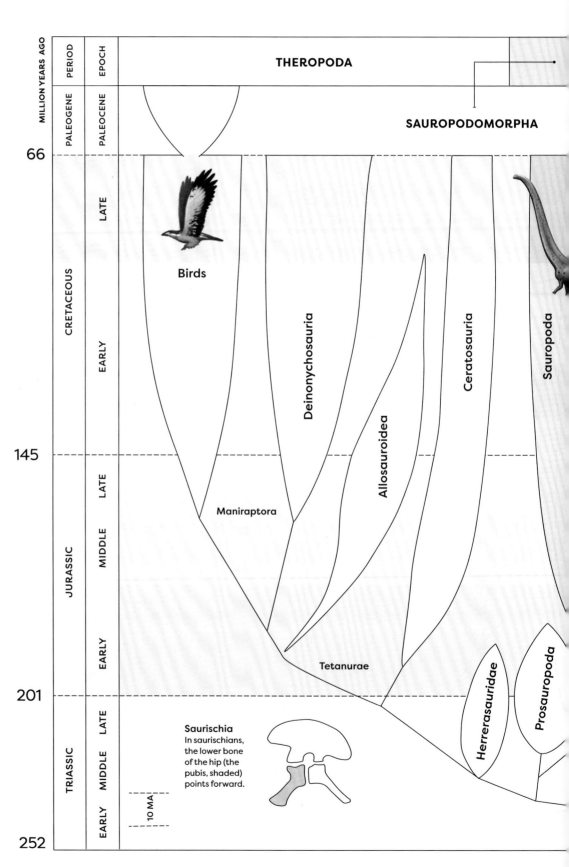

Saurischia
In saurischians, the lower bone of the hip (the pubis, shaded) points forward.

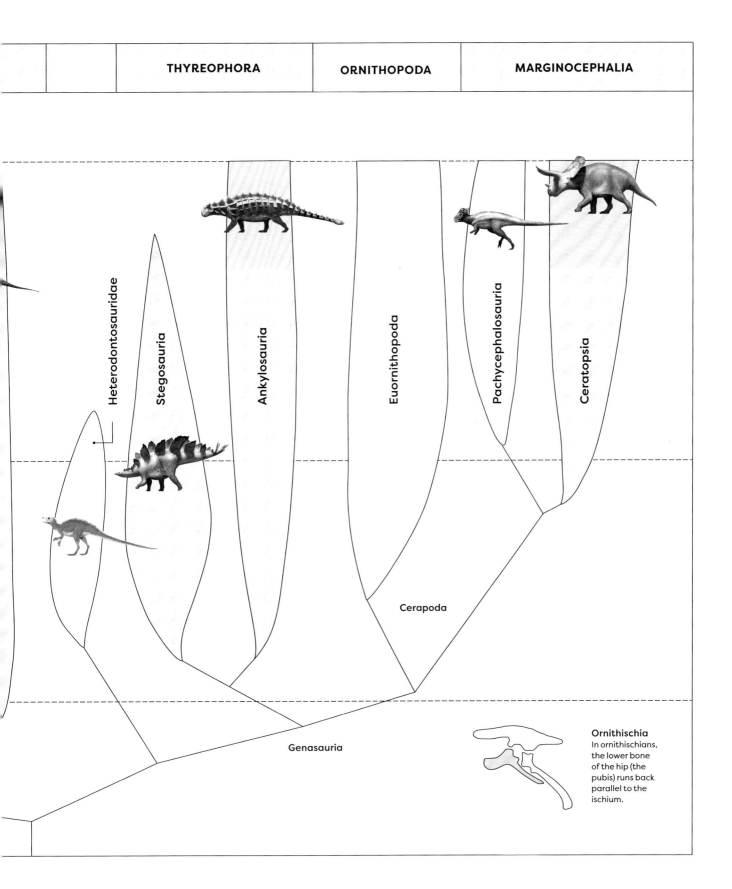

Ornithischia
In ornithischians, the lower bone of the hip (the pubis) runs back parallel to the ischium.

BIRDS AS A BRANCH OF THEROPODS

It's wonderful to know that birds are dinosaurs and you can see them every day; but birds arose in a remarkable way through the Jurassic period that involved a reduction in their body size.

When the first skeleton of the earliest bird *Archaeopteryx* was found in southern Germany in 1861, scientists realized this was a clue to the origin of birds. The fossil was clearly a bird, about the size and shape of a pigeon and with definite feathers preserved in the rock. But it had the skeleton of a small theropod.

For a long time, *Archaeopteryx* was the only fossil bird known from the Jurassic or Cretaceous—it had so many bird features, including feathers, wings, a fused clavicle (wishbone), hollow bones (to save weight), special wrist bones to enable it to fold its hand back, and big eyes and brain so it could see in three dimensions. Did this mean that birds had evolved really fast? The absence of older and younger bird fossils meant that *Archaeopteryx* became the subject of debate about rates of evolution, and even about whether evolution could explain the origin of such an amazing group as birds.

Today there are 10,000 species of birds, and these include many amazing flyers. How did this astonishing package of adaptations originate? Thanks to thousands of beautiful fossils from China, we now know much more than we did twenty years ago. These fossils include dinosaurs close to being birds and many other early birds from the Jurassic and Cretaceous. Three amazing facts are now known for sure: 1) Feathers did not arise with the origin of birds but much earlier, probably when dinosaurs arose in the Early Triassic; 2) Some dinosaurs could fly; 3) Birds and flight were possible because one lineage of theropods became smaller and smaller through the Jurassic, a process called miniaturization.

Let's look at *Microraptor*. This little dinosaur, a relative of *Deinonychus* from the United States, had birdlike flight feathers on its arms and legs, and recent calculations show

THE WORLD'S FIRST BIRD
This *Archaeopteryx*, from the Late Jurassic of Germany, is one of fifteen or so specimens, and it is kept in the Humboldt Museum in Berlin, Germany. Notice the dinosaur-like skeleton, but the modern, birdy feathers over its body, legs, and wings.

it could fly, flapping all four of its wings. How do we know it had feathers? The fossils show them clearly, and these allow aerodynamics experts to work out its effective wing area, and it turns out it could flap its wings and power through the air. We even know it had black, iridescent feathers.

So, not only are birds dinosaurs, and they evolved all their special flight adaptations over 50 million years before *Archaeopteryx*, but we now even know that some bird cousins such as *Microraptor* could fly, and this was a different kind of flight than seen in *Archaeopteryx* and in birds in general.

MINIATURIZATION

It's not common for animals to become smaller through time. In fact, the opposite is more usually the case, and this is so common that it has its own name, Cope's Rule. This idea was named after Edward Cope (1840–97), one of the greatest dinosaur paleontologists of the 1800s. He noticed that dinosaurs started small and grew bigger through time. It was the same for horses: the first horses were about the size of a terrier dog, and through 50 million years they became larger, up to the size of modern horses.

It's easy perhaps to see why Cope's Rule is so common in animal evolution. Being big has many advantages, such as being able to dominate your landscape. If you are a herbivore, you can get all the food and escape being attacked, such as modern elephants, and if you are a carnivore, being large allows you to attack anything. There are costs, of course, in being large, such as that you need more food, it takes longer to reach breeding age, and your babies take forever to grow up.

OPPOSITE
The skeleton of *Microraptor* shows all the features of a maniraptoran dinosaur, but there are lines of elaborate flight feathers along the arm and leg. In the fossil, the dark tips of the feathers show up against the rock.

IN FLIGHT
Microraptor, a dromaeosaurid like *Deinonychus*, had flight-type feathers along its arms and legs, and aerodynamic calculations show it could take off and fly by flapping its front and back wings. So, some dinosaurs could fly—it wasn't just the birds!

So, while theropods such as *T. rex*, *Giganotosaurus*, *Carcharodontosaurus*, and *Spinosaurus* were becoming giants, each weighing about 5 tons, one group, the maniraptorans, were getting smaller. They started in the Triassic at about 15 ft (5 m) long and weighing 260 lb (120 kg), and over 50 million years through the Jurassic became smaller and smaller, eventually reaching chicken size weighing about 1 lb (0.5 kg).

At the same time, the arms and hands of the maniraptorans were getting longer and stronger. The name "maniraptoran" means hand-hunter, and they had long powerful fingers for grasping prey. Also, we now know from the Chinese fossils, their arms were lined with evenly spaced flight feathers, so from a certain point in the Jurassic they were capable of gliding, perhaps leaping from tree to tree in search of juicy insect prey. At a certain point, the balance of reducing body weight and increasing arm length and wing size meant they crossed the crucial point of take off in powered flight.

As we shall see (page 98), there is a particular point at which a bird or an airplane can take off, which requires the correct ration between weight and wing area. Reducing weight (through miniaturization) is a great way to reach that point.

DINOSAURS IN PERSPECTIVE

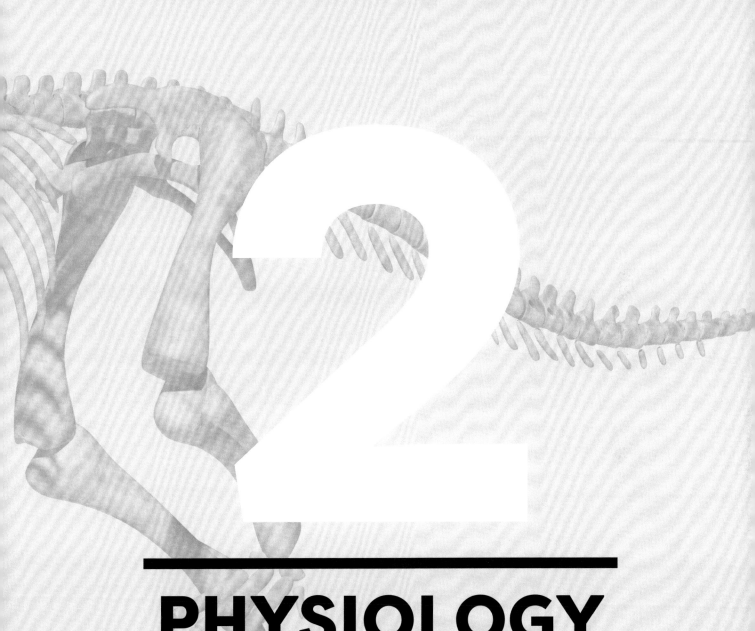

2
PHYSIOLOGY

VISUALIZING DINOSAURS

We have seen dinosaurs recreated in many different ways over the past two hundred years, and now, with new scientific understanding there are some amazingly realistic models, artwork, and even movies.

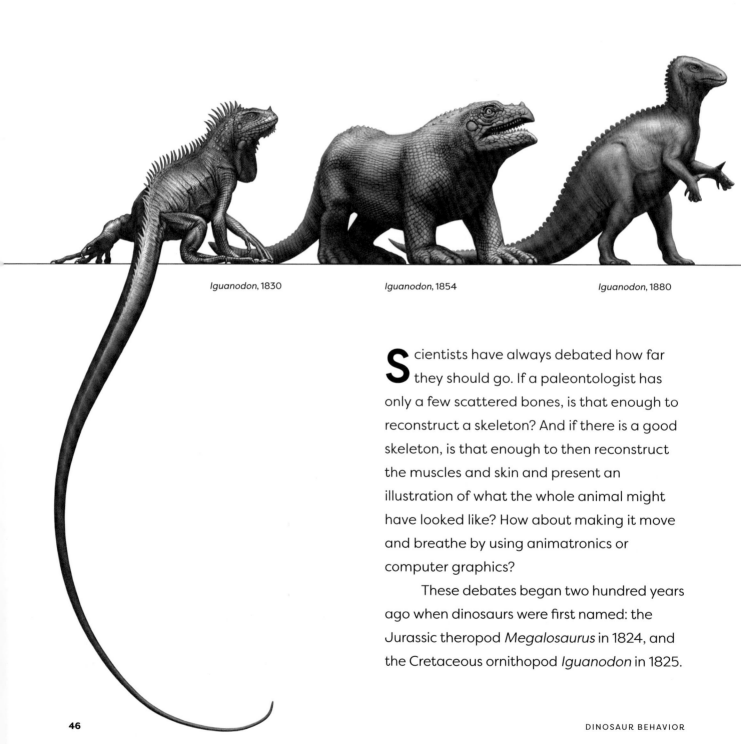

Iguanodon, 1830

Iguanodon, 1854

Iguanodon, 1880

Scientists have always debated how far they should go. If a paleontologist has only a few scattered bones, is that enough to reconstruct a skeleton? And if there is a good skeleton, is that enough to then reconstruct the muscles and skin and present an illustration of what the whole animal might have looked like? How about making it move and breathe by using animatronics or computer graphics?

These debates began two hundred years ago when dinosaurs were first named: the Jurassic theropod *Megalosaurus* in 1824, and the Cretaceous ornithopod *Iguanodon* in 1825.

Iguanodon, modern-day

DINOSAUR IMAGE EVOLUTION
Here, the sequence from left to right shows *Iguanodon* first as something like a giant *Iguana* lizard, in about 1830, and then as it was reconstructed in 1854 for a life-sized model on show in London. Then, after complete skeletons were found, we see an upright, bipedal stance, something like a kangaroo. The modern reconstruction shows a more balanced biped, sometimes walking on all fours.

Was the job of a paleontologist simply to illustrate the fossil bones and not to make any guesses about what the animal looked like when it was alive, and even what it ate and how it behaved? Or, on the other hand, should the paleontologist use all the clues available to show people the best understanding they had of a lifelike dinosaur image or model?

The debate continues today, with some paleontologists seeing their job as simply presenting the science and not speculating or guessing about anything. They object to museum displays with lifelike paintings or animated dinosaur models. Other scientists believe they should share their work with the wider public, using models, paintings, and computer animation to do so.

Paleontologists attempted the first drawings of "living" dinosaurs in the 1820s, but the first reconstructions to hit the headlines were the famous Crystal Palace models unveiled in 1854. These were life-sized models made from steel and concrete by the artist Benjamin Waterhouse Hawkins (1807–94), and they were a huge hit with the public in London. Later, Charles R. Knight (1874–1953) produced the first modern-style paintings of all the newly discovered dinosaurs in the American West, and

PHYSIOLOGY

OPPOSITE TOP
Two dinosaurs as reconstructed in 1854 for the Crystal Palace display by Waterhouse Hawkins. Here, two *Iguanodon* look fierce and scaly, with horns on their noses (later identified as their thumb claw).

OPPOSITE BELOW
A long-necked plesiosaur sits on the edge of the lake, which represents the sea. This reptile lived in the Jurassic oceans, where it hunted fish, swimming by beating its large front paddles.

BELOW
Two leathery-winged pterosaurs stand on a rocky cliff, planning their next move. Hawkins shows their long beaks, lined with sharp little teeth, ideal for trapping fish. Their wings are folded, but the pterosaur behind is extending its wings ready to take off.

his images were seen in the leading museums as well as in magazines and popular books.

Everything changed in the 1990s when artists could use computer-generated imagery (CGI) in movies such as *Jurassic Park* (1993) and the BBC documentary series *Walking with Dinosaurs* (1997). But times change. These early CGI films show dinosaurs with scaly skin, whereas we now know that many dinosaurs had feathers (see page 58). It's hard to know when we will be fully confident that our image of dinosaurs is finally correct and accurate, but paleontologists think we are getting there.

TOP
The famous scene in the first *Jurassic Park* movie in which an escaped *T. rex* menaces the hero. Here, the dinosaur is purely scaly, but the moviemakers took care in showing a slow running speed. Now, we would probably show *T. rex* with some feathers.

LEFT
The intrepid explorers of *The Lost World* see a *Stegosaurus* feeding on ferns and tree leaves. It's shown as active, with a high tail, not trailing along the ground as in older reconstructions.

DINOSAUR BEHAVIOR

FORENSICS: PUTTING FLESH ON THE BONES

It's not just new digital technology that creates realistic-looking modern dinosaur illustrations; it is also our ever-growing knowledge from fossils. The recent dinosaur discoveries in China revolutionized everything. Before the 1990s, things were simple: birds have feathers, dinosaurs don't. Then, paleontologists in China began unearthing more and more dinosaurs with feathers (see page 58): first *Sinosauropteryx* in 1996, then *Caudipteryx* in 1998, and *Microraptor* in 2000. These three discoveries showed that many small theropods had feathers, and nobody knew where the new finds would end. Discoveries in 2002 showed that *Psittacosaurus* had feathers too, which was difficult to understand as this dinosaur was located far from birds in the evolutionary tree. Could it be that all dinosaurs had feathers?

Feathers give previously scaly dinosaurs a whole new look, and in 2010 paleontologists started to be able to identify the colors of dinosaur feathers (see page 172). This meant that paleoartists—those who make paintings and models of dinosaurs—sat up and paid attention. Everything they had done up to 2010 might now be wrong. No more scales, no more dull colors!

The image shown below of a *Sinosauropteryx* by paleoartist Bob Nicholls shows all the details of its feathers and their color patterns—and we are confident about these details because of remarkable fossil preservation. *Sinosauropteryx* was covered all over with short, bristly feathers, even on the tail and legs and over the face (see page 171). The fossils from China (see pages 60–1) show all the bones and feathers, and it is possible to identify the colors as ginger and white (see pages 172–3), and even the dark-colored "bandit mask" around the eyes.

Bob created this image digitally, using an art app, so he could generate each feather and scale as a separate element and fix the colors. New data can be used to edit the detail at any time, so the image is dynamic. It can be turned from a flat, two-dimensional image into a three-dimensional image for animation or display. Welcome to the new age of paleoart!

FLESHED OUT
From scattered bones (left) to living dinosaur (right). Here we see the sequence of fleshing out *Sinosauropteryx*, a feathered maniraptoran dinosaur. Its scattered tail bones are fitted together to make the complete skeleton in a lifelike pose, muscles are added, then skin, then feathers, all based on evidence from fossils.

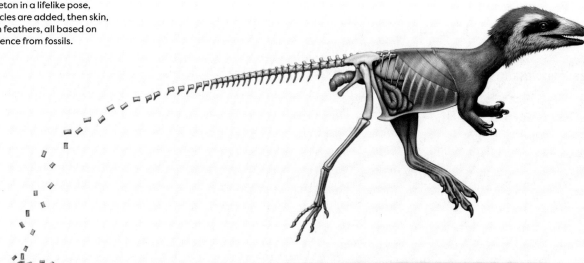

PHYSIOLOGY

MODERN REPTILES AND BIRDS

Understanding the physiology of modern reptiles and birds helps us to understand dinosaurs, and there are huge, significant differences between the functioning of the two.

At one time, paleontologists concentrated on modern reptiles such as lizards and crocodiles when they were reconstructing dinosaurs. The idea was that the dinosaurs were cold-blooded and slow-moving, just like an alligator in the zoo. The alligator enjoys lying out in the sun, absorbing the heat from its rays, and it moves only from time to time, and those movements are slow. It may wander over to the water and slide in for a lazy swim, or crawl on its belly toward a hunk of meat left by the zookeeper.

This is the classic depiction of a cold-blooded reptile. Lizards and crocodiles are ectothermic, meaning "external-heating." They get most of their warmth from the environment,

DAILY BODY TEMPERATURE

Warm-blooded animals (endotherms) show almost constant body temperature all day, whereas cold-blooded animals (ectotherms) can show body temperatures that vary up and down in line with changing air temperatures in the day and at night.

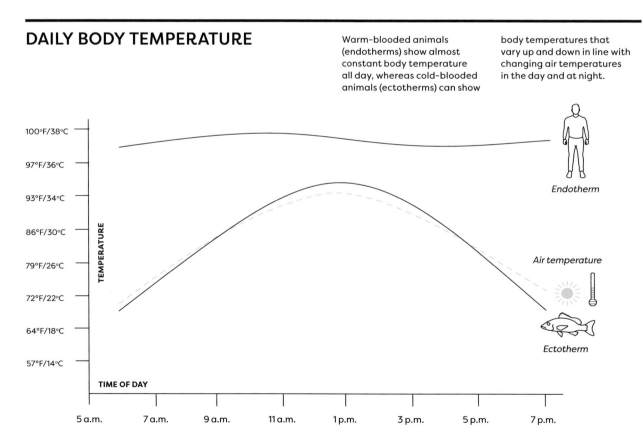

ARE ENDOTHERMS BETTER THAN ECTOTHERMS? This (warm-blooded) egret might say so as it takes its perch on the back of a (cold-blooded) alligator. Different strategies for different modes of life.

by basking on rocks and absorbing heat from above and below. Small cold-blooded reptiles can run fast to escape, but they shoot under a rock as soon as they can because their energy levels are low and they cannot run far (see page 88). A big advantage for ectotherms, which prefer to live in warm climates, is that they don't need to eat much food.

Endotherms ("internal-heaters"), on the other hand, such as birds and mammals, have to eat plenty. An endotherm of the same body mass as an alligator, such as an ostrich or a human, has to eat ten times as much food each day because 90 percent of its diet is devoted to firing up the inner furnaces. Advantages for endotherms are that they can live all over the world, even in cold climates, because they generate their own heat and keep their body temperature constant. They can also hunt at night when the air is cold and other animals may be less alert. A constant body temperature means that all the chemical reactions inside the body are tuned to that temperature and can work highly efficiently.

These differences between ectotherms and endotherms are key parts of their physiology. "Physiology" refers to all the chemical processes in the body relating to feeding, breathing, energy transfer, water transfer, and use of the muscles. A key part of physiology is the metabolic rate of an animal: this is the rate at which oxygen is consumed. Ectotherms breathe at a low rate when they are not alarmed, whereas endotherms have higher rates of processing oxygen due to their faster, active, higher-intensity lifestyles.

So, do we see dinosaurs as lumbering around slowly and sleeping most of the time? Or do we see them as active, fast moving, and capable of complex behavior?

A SPECIAL KIND OF WARM-BLOODEDNESS

Cold-blooded or warm-blooded? The debate about dinosaur temperature regulation hugely affects how we interpret all aspects of their behavior.

Were dinosaurs ectotherms or endotherms? Up to about 1970, paleontologists were convinced they were ectotherms, but then evidence from detailed bone structure showed we should reinterpret them as endothermic. The bones show all kinds of evidence for this. First, the internal detail shows generally continuous growth and an open structure, as seen today in the bones of mammals and birds. Second, the bones show an overall microscopic structure typically seen today in birds and mammals (see "Bone histology," opposite). A third line of evidence in the bones of dinosaurs came to light in 2022, when Jasmina Wiemann of Yale University showed that nearly all dinosaurs were endothermic. She found high levels of particular organic waste chemicals in bones of birds, mammals, and dinosaurs, matching their high use of food, but these were absent in cold-blooded lizards and crocodiles.

A key point about many dinosaurs is that they were huge, and this gave them an automatic means by which to control variations in their body temperature. A small lizard, for example, warms up in daylight and cools down at night. An alligator warms and cools more slowly because of its size, and a dinosaur would have warmed and cooled yet more slowly. This gives an effect called mass homeothermy, meaning steady body temperature because of huge size.

Mass homeothermy is a result of the surface-area-to-weight relationship. Small animals have lots of skin over their bodies relative to their mass, whereas the relative area in large animals is much smaller. This is why large endotherms such as elephants and whales need less food and oxygen per ton of body mass than smaller animals. Tiny endotherms such as mice and humming birds have to eat huge amounts and breathe superfast to take in enough oxygen to counter the loss of body heat from their tiny bodies.

So, a giant dinosaur such as *Brontosaurus*, which weighed ten times as much as an elephant and was also endothermic, probably did not have to eat ten times as much food as an elephant. It could rely on the natural properties of its huge body to retain core heat without any effort.

BONE HISTOLOGY

The study of cellular structures in the body is called histology, and scientists have long used the microscope as a means to learn more. Biologists and physicians study sections of the skin, bone, and internal organs of modern animals and of humans, to recognize different cell types and indications of disease.

Humans are each made up of 37 trillion cells, but these can be classified into about two hundred different types, each specialized for a different function. For example, there are several different kinds of nerve cells, a variety of cell types involved in the bloodstream, several different types in the muscles, and of course numerous cell types in bone.

REACHING FOR THE LEAVES
The giant sauropod *Brontosaurus* rears up for a moment to grab a tasty mouthful of leaves. The large dinosaurs had constant body temperatures, mainly because they were so huge.

PHYSIOLOGY

Bone is a living tissue, made from nerves, blood vessels, fat, flexible cartilage, and needle-shaped crystals of calcium phosphate (apatite) that form the hard structure. Living bone is partly flexible and partly brittle, and the degree of flexibility depends on age. For example, if a child falls over, they may be bruised and their bones may bend a bit then bounce back. If the child receives a harder knock, a bone may break, but provided the broken ends are realigned and the limb is bound up tightly with a splint, the bone will knit back together. In older people, the bones are less flexible and break more readily.

Bone structure differs between ectotherms and endotherms, and these differences can be seen in fossil bones. Paleontologists were amazed, over a hundred years ago, when they began to look at microscopic sections of fossil bone. Just as biologists take thin sections of tissues from modern animals, paleontologists can take a fossil bone and make a thin section—a very thin slice, thin enough to allow light to pass through but showing enough of the structure. Under the microscope, paleontologists saw that the detail of fossil bone could be as good as modern bone. Everything was there, and all that had changed was that the spaces filled with soft tissues such as blood vessels and fat were filled with minerals.

BODY SIZE AND METABOLIC RATE

Metabolic rate, measured in terms of oxygen consumption, increases with body size. Today, reptiles fall on a lower line than mammals because they consume less and require less oxygen because they are cold-blooded. Estimated values for dinosaurs fall high in the graph, much more in line with mammals than modern reptiles, because they too were probably warm-blooded.

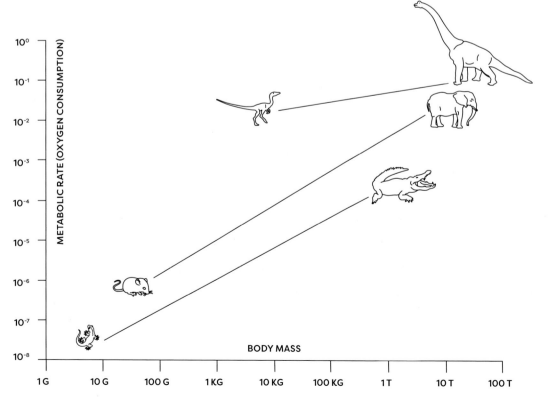

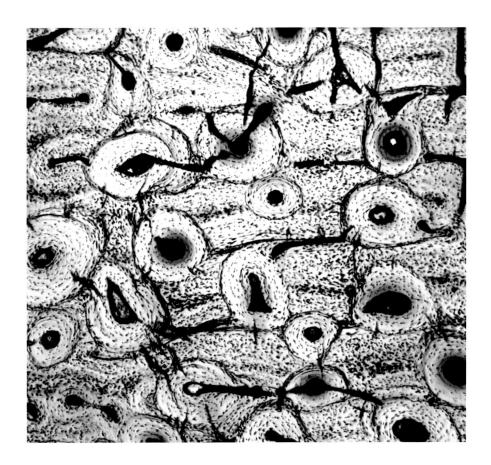

SLICE OF BONE
A thin section (slice) through dinosaur bone, viewed under the microscope. It shows the so-called fibrolamellar structure throughout, composed of layers of pale-colored bone and showing darker spaces through which blood vessels and nerves ran. This kind of bone is associated with fast growth and warm-bloodedness. Superimposed on the background structure are larger circular structures called Haversian systems, which indicate sites where minerals have been eroded from the bone to fuel the dinosaur's growth. New bone has then been deposited in rings inside the tubes, filling inward and leaving a circular black space at the center where the blood vessel passes through.

In studies of dinosaur bone in the 1970s, paleontologists showed that it was usually of a type called fibrolamellar, meaning it was composed of regular, interconnected, open, layered structures. This is typical of the bone of living endotherms such as birds and mammals that experience fast growth.

Another feature of endotherms is called secondary Haversian remodeling. This is where the bone has been reworked, meaning that the original bone has been dissolved in places to acquire minerals into the bloodstream, then new bone has been laid down, creating overlapping circular structures called secondary Haversian systems. These systems are seen usually only in animals with high metabolic rates, where minerals such as calcium and phosphorus are needed in rich supply and fast, when the animal is particularly active.

Modern reptiles such as turtles, lizards, and crocodiles, on the other hand, typically do not develop these bone tissues, or not so frequently. Instead, they have a bone type called lamellar-zonal bone, showing cycles of growth. Also, they lack secondary Haversian remodeling. Both of these features suggest slower growth and lower metabolic rates, typical of a cold-blooded, or ectothermic, animal.

So, the high quality of fossils makes it possible to determine metabolic rates of dinosaurs from their bones, and it is clear that they were endothermic, or warm-blooded.

FEATHERS

Birds have feathers, but so did most dinosaurs. This is a new and still controversial idea, but the evidence appears to be clear, and it matches with their warm-bloodedness.

Older images always showed dinosaurs with scaly skins, like giant crocodiles or snakes. Indeed, many dinosaurs did have scales or bony plates in their skin, sometimes as a form of armor or simply to protect them from injury. In the 1970s, some researchers speculated that dinosaurs might have had feathers, but it was a discovery in 1996 that answered the question.

This was the year when the first feathered dinosaur, *Sinosauropteryx*, was announced from the Cretaceous of China. At first, some paleontologists did not believe it and said the simple, bristly feathers were bits of shredded scales or muscle. But over the following years, as more and more feathered dinosaurs were dug up in China, people had to accept the evidence: feathers were not unique to birds but were found in many dinosaurs, too.

What were these feathers for? In birds, feathers have three main functions: to insulate the body and keep it warm; specialized feathers for flight; and for display. In 2010 a way was found to tell the color of dinosaur

FEATHERS AND SCALES
The plant-eating dinosaur *Kulindadromeus* from the Jurassic of Russia shows an amazing mix of feathers and scales of different types over its body. Fossils have shown all this detail, including the broad armor-like scales down the tail, smaller scales over the legs, and tight feather covering elsewhere.

FEATHER TYPES
The seven types of feathers found in modern birds, from the pennaceous feathers of wing and tail on the left to fluffy down, contour, and semiplumes, and whiskery bristles and filoplumes on the right.

feathers (see page 172), and this showed they were colorful and probably used for display.

Then, in 2014 the further shocking discovery of an ornithischian dinosaur called *Kulindadromeus* with feathers meant that probably all dinosaurs had feathers, not just the theropods close to the origin of birds. Even more startling, in 2018 and again in 2022 evidence was published to show that the pterosaurs, the flying reptiles that are close cousins of dinosaurs, also had feathers. Pterosaurs and dinosaurs branched apart in the evolutionary tree back in the Early Triassic, (see page 35), so this suggests that the very first dinosaurs and their immediate ancestors also had feathers. Feathers perhaps first arose to provide insulation, then they became colorful and patterned for display, and finally the pennaceous feathers arose in theropods for gliding and flight.

ALL KINDS OF FEATHERS

There are many kinds of feathers, including simple whiskers, fluffy down, and pennaceous feathers. These are the feathers with a quill and branching barbs on each side, and they include the flight feathers in the wing as well as covert feathers over the shoulders, strengthening the wings, and on the tail.

In fact, modern birds have seven types of feathers, and these can be called generally either filaments or pennaceous feathers. The filaments include three forms. Bristles are stiff, and they are found around the eyes and face, providing some protection. Second are filoplumes, which are also stiff but have some small side branches called barbs, and have a sensory function—they are often located around the bird's beak and can be used to feel for prey such as insects that the bird wants to snap up. The third kind of filaments are down feathers, which are fluffy structures with many branches coming from the root, and they provide insulation, covering the bird's body.

The other four feather types are pennaceous, meaning they have a central quill, or rachis, and numerous side branches, or barbs. These are all larger feathers that form most of the wings and tail, and cover over the back. The first type is the semiplume, a feather with numerous barbs, but the barbs do not interlock, so these feathers are also fluffy, like

PHYSIOLOGY

down, and are mainly for insulation. Then, the contour feathers have barbs with tiny hooklets that can interlock and hold the shape of the feather. Contour feathers cover the top of a bird's wing and also extend over the back.

The sixth and seventh feather types are what people usually think about when they pick up a discarded feather from a pigeon or seagull. These are the tail feathers (retrices) and wing feathers (remiges), often long, with a rachis and lateral barbs, which form complete surfaces for flight. In these feathers, it is important that the whole structure is tough and air-tight. If, for example, either tail or flight feathers allowed air through, they would not function so effectively as solid surfaces.

We can see how important those feathers are when we watch birds. They often stretch out a wing and run their beak through the feathers, called preening. Preening is essential for removing seeds and parasites from the feathers, but also for lining up the barbs and removing gaps and knots between them.

DINOSAUR FEATHERS

When paleontologists began to identify feathers in dinosaurs such as *Sinosauropteryx*, *Caudipteryx*, and *Microraptor*, they seemed to fit the seven types of feather seen in modern birds. But then, something strange was noticed: some dinosaurs had feathers that were unlike anything seen in a modern dinosaur. For example, when Xu Xing and colleagues studied the oviraptorosaur theropod *Similicaudipteryx*, they found it had strange ribbon-like feathers, with a long, flat rachis, and a pennaceous tuft of branching barbs at the far end, like a flag. Looking closely at the feathers of other dinosaurs, they identified some other strange forms. For example, the Early Cretaceous ornithischian *Psittacosaurus* (see page 174) has a closely packed row of cylindrical bristles in a line along the middle of the tail; each feather is over 6 inches (16 cm) long, stiff, and standing upright.

The Middle Jurassic ornithopod *Kulindadromeus* from eastern Russia (see page 58) has a great range of types of scales and feathers, including bristles around the head and body, and pennant feathers all over the body—these have a circular basal plate from which five to seven slender filaments trail backward. These particular feathers are somehow halfway between scales and feathers, and indeed modern birds can have both. For example, a chicken has feathers all over its body, but its legs are scaly like a reptile. *Kulindadromeus* also had scaly legs, and a scaly tail. So it seems feathers and scales could be interchangeable in dinosaurs and birds.

Recent studies of pterosaurs, the cousins of dinosaurs (see page 24), show that they had feathers too, and at least four types: simple bristles, bristles with a tuft at the end, bristles with a tuft halfway down, and down feathers.

It may be unexpected or shocking that dinosaurs and pterosaurs had feathers, and that they showed a greater variety of feather types than we see in modern birds. However, these Mesozoic reptiles existed for over 160 million years, and their feathers probably had many functions, so perhaps it is not surprising after all that so many amazing types of feathers evolved.

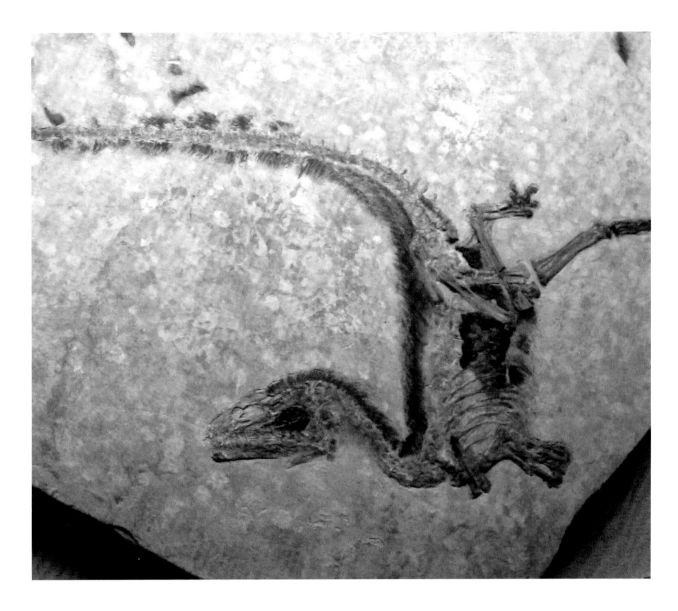

THE FOSSIL THAT CHANGED EVERYTHING
When this specimen of *Sinosauropteryx* was published in 1996, it was the first time that feathers were observed in something that was not a bird. Notice the bristle-like, short feathers over the back of the head, along the body, and in tufts down the tail.

PRESERVED IN AMBER

Amber is a remarkable substance: a transparent yellow-orange mineral formed by solidification of tree resin. Trees, especially conifers such as pine or spruce, produce glue-like resin to help heal breaks in their bark and other damage. Amber can be used as a semiprecious jewel, and pieces are sold to be worn as pendants, necklaces, or rings. Especially popular over the centuries have been amber pieces that contain a fossil insect. The insect may be well preserved, showing beautiful detail such as the fine whiskers on its

PHYSIOLOGY

LEFT AND RIGHT
The most amazing dinosaur fossil ever? This tail of a small dinosaur is locked in amber and shows every detail of the long, branching feathers. These are definitely feathers, but there is nothing like them in any living bird. The whole specimen (right) would sit neatly in the palm of your hand (can you spot the ants?). In a close-up of the same specimen (left), you can see the tips of the long, whiskery pennaceous feathers—a type not seen in any bird.

legs, or beetles can show some of their original colors, for example.

Paleontologists have studied amber fossils for a long time, from specimens found in north Germany, the Dominican Republic in the Caribbean, and from Burma (Myanmar) in Southeast Asia. The amber from Burma has been especially exciting recently because some amazing specimens have been found in it, including tiny lizards, frogs, and even birds.

Burmese amber is Cretaceous in age, about 100 million years old, and it samples a huge variety of amazing tiny plants and animals. This was when new kinds of plants, the flowering plants or angiosperms, were appearing on Earth, and the new flowers attracted all kinds of new insects, which in turn were fed on by all kinds of new lizards, birds, and mammals.

Paleontologists used to joke about the "dream fossil": imagine finding a *dinosaur* in amber! But this seemed a ridiculous hope because dinosaurs are big and amber usually preserves only small animals. Then, in 2016 the miracle fossil was found. Well, in fact it was just a part of the tail of a very small dinosaur. However, this specimen of clear, pale yellow amber, a piece that would sit easily on the palm of your hand, shows a fluffy tail, with long, slender feathers. At first, paleontologists thought the fluff was made from hair and that this maybe came from an early mammal, but in close-up the long fluffy structures show numerous side barbs. They are feathers. How did paleontologists know it wasn't a long bird tail? Well, on X-ray-scanning the fossil, they found the skin, muscles, and bone preserved inside, and the bones were the tail bones of a dinosaur, not a bird.

BREATHING

Dinosaur breathing may have been super-efficient as in birds, and this was a smart way for them to be large and active at the same time.

Breathing is such a regular activity we don't usually think about it. We breathe in to take oxygen into the body, and we breathe out to expel the waste carbon dioxide. We know that our lungs process the gases, passing oxygen into the bloodstream, where it is carried around attached to red blood cells. After the oxygen has passed into the muscles and gut tissues, carbon dioxide returns into the bloodstream and back to the lungs. The oxygen passing through the body travels in the blood vessels called arteries, and the oxygenated blood is bright red. The returning blood is darker in color and passes through the veins and back to the lungs. The whole system is, of course, powered by the heart which pumps the blood through four chambers, allowing for so-called "double circulation," an efficient way to avoid mixing red oxygen-rich blood with the darker venous blood.

The system of breathing that mammals such as humans have is not entirely efficient. The problem is that we breathe in and out—a so-called tidal system—and when we breathe out, we don't entirely expel all the waste air. A volume remains in the lungs and breathing passages, and we pump in the next mouthful before clearing the old air. Birds, on the other hand, have a straight-through system, involving multiple air sacs as well as the lungs, and air goes in one way and out the other, so there is no dead space.

Dinosaurs almost certainly had the same system. There are no fossil lungs of dinosaurs, but there is evidence for the additional air sacs, shown by openings and hollow spaces in the vertebrae, the bones of the backbone, as well as in ribs and limb bones. This is called pneumatization, the invasion of bones by air sacs, both to save weight and to accommodate the air sacs. Weight-saving is important for birds to enable them to fly, but it's also important for large dinosaurs, to save weight. This is especially true of the long-necked sauropods, for example; with their air sacs, the neck weighs about half as much as it would if the bone were solid throughout.

The birdlike one-way breathing system in dinosaurs helped them to raise their metabolic rates and be endothermic, but at less cost than if they had had a tidal breathing system.

This is probably part of the smart set of adaptations they had to balance the costs of being large against the need to be able to move about reasonably actively, and without having to eat crazy amounts of food.

PNEUMATIC BONES

Modern birds and crocodiles have spaces inside their bones that in life are filled with air or fluid. The pneumatization of skull bones in crocodiles is probably a holdover from more ancient times when they perhaps had different modes of life or different respiration systems. Mammals even have some pneumatization; for example, humans have air sinuses inside the skull, above and below the eyes, and these spaces can become painful when you have a bad cold.

In birds the pneumatization is a key part of their metabolic system. Nearly all bird bones are hollow, but the hollow leg bones, for example, contain bone marrow, the spongy tissue in the body that produces red blood cells. The other hollow bones, in the vertebrae of the neck, trunk, and hip region, as well as their ribs, breastbone, and upper arm bones, are occupied by air sacs, balloon-like structures with an outer membrane and filled with air in life.

Why have such a complicated system? First, hollow bones save weight, and that's important for a flying bird as its success depends on flight ability. Air spaces can also help to rebalance an animal, by shifting weight back to the hips and tail, for example. Some birds even use their ability to pump air in and out of the air sacs as a way to survive when they fly up to great heights where air pressure is lower than near the ground.

Dinosaurs did not have pneumatization for that reason, but weight-saving and rebalancing did matter. It was no surprise when paleontologists discovered that

---

BREATHING TYPES

TIDAL BREATHING
In tidal breathing, as in humans and all mammals, the air goes in and out.

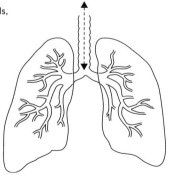

LIKE THE MOTION OF OCEAN TIDES

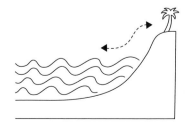

UNIDIRECTIONAL BREATHING
The air breathed in passes through a circuit, giving oxygen to the body and picking up carbon dioxide to breathe out.

LIKE THE MOTION ON A RACETRACK

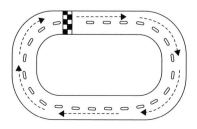

PHYSIOLOGY

sauropod dinosaurs had extensive pneumatization throughout nearly all of the vertebral column, from head to tail, as well as in their ribs and upper arm bones. It has been estimated that pneumatization makes the sauropod neck about half as heavy than if it were made from solid bone. And that matters if you have a 30-ft (10-m) long neck that has to be held up by huge muscles coming from the shoulder area.

What about the smaller theropods? Nearly all theropods, just like their descendants the birds, had hollow bones. In this case, it's a little about weight-saving for the larger examples such as *T. rex* and *Spinosaurus*. But it's as much for adjusting their balance. A typical large theropod saves weight in its forequarters, so its center of mass shifts backward to sit just above or in front of the hips. Theropods were all bipeds, walking just on their hind limbs, so keeping balance mattered, and it's been worked out that shifting the center of mass back a little would have increased their agility—especially their ability to twist and turn when chasing prey. And of course, it's also about improving their breathing systems. Dinosaurs shared the same one-way breathing system seen in birds, and that maybe gave them a 10 percent improvement in efficiency in transporting oxygen into the body and carbon dioxide out.

RESPIRATORY INFECTION

A huge *Brontosaurus* lumbers into view in the Late Jurassic landscape of tall conifer trees. It's a misty daybreak, with cold clouds of water vapor gathering over the lakes of the Morrison Formation scene. The *Brontosaurus* sneezes and snuffles. It lets rip with the largest sneeze ever. A small herd of browsing *Stegosaurus* look up nervously and shift further into the forest. The *Brontosaurus* plods on, breathing in and out with difficulty, making great roaring and rasping noises as it sucks air down into its lungs, and puffs it out again from the air sacs around its neck and torso. Great gobbets of snot fly from its nostrils as it puffs out a labored breath. Its head droops miserably. Here is a sick dinosaur.

How do we know dinosaurs suffered from respiratory diseases? This comes from a 2022 investigation by Cary Woodruff of the Great Plains Museum and Ewan Wolff at the University of New Mexico. They were studying a skeleton of *Brontosaurus* in the Museum of the Rockies, Montana, nicknamed Dolly. We have no idea whether Dolly was female or male, but the researchers argued they have clear evidence that this particular dinosaur was suffering from serious respiratory disease when it died.

The evidence comes from lesions, or unusual bone growths, on the vertebrae of Dolly's neck, just at the point where the air sacs entered the sides of the vertebrae. The bone lesions are irregular growths about 1 inch (2.5 cm) across on an otherwise flat structure, showing infection of the bone. The infection caused the bone to react while the animal was alive, growing unusually fast on the surface to try to respond to the attacking infections. In X-ray sections through the bone, the investigators found evidence of unusual and excessive bone growth in and below the lesions.

What could have caused the infections? It might have been a fungus, and such

AAAHH...CHOO!
The biggest sneeze ever is made by this Jurassic *Apatosaurus*. It has a serious throat infection, similar to those found in modern birds.

infections are seen in modern birds, termed aspergillosis. The fungal spores are microscopic and can enter the respiratory system through the nose or mouth, where the warm, damp environment encourages their growth. They then cause irritation within the breathing system, and a common response is that the membranes become red and infected, and they even attack the bone, which grows fast, attempting to overgrow or swamp the irritant. In modern birds, aspergillosis causes labored breathing, wheezing, sneezing, and over-production of mucus.

Dinosaurs didn't have vets, so they could not receive any medicine. However, aspergillosis is also a common disease of chickens and there is no cure, so Mesozoic medical aid would probably not have helped poor Dolly.

FOOD BUDGETS

In many cases we know what dinosaurs ate, but working out how much they ate each day is difficult.

As we will see later (see page 70), paleontologists use all sorts of evidence to identify what particular food dinosaurs ate. The easiest thing to determine is whether they ate plants or meat, and then to compare all the dinosaurs and other beasts that lived together to work out a food web. A food web is a diagram that identifies who eats what.

Dinosaurs were parts of wider ecosystems; in other words, the whole network of species of plants and animals that lived together in a particular setting, and how they interact with one another. In many cases, paleontologists know a great deal about all the plants and animals that lived together, and the food web diagram may include different kinds of plants, pond life such as snails, shrimps, and fish, as well as land-dwellers such as lizards, crocodiles, pterosaurs, birds, mammals, and the dinosaurs.

Understanding something about a modern or ancient ecosystem allows biologists to understand how energy flows. Energy enters the ecosystem from the sun, and it is captured by plants that take in the sun's energy and convert carbon dioxide from the air and water that they draw up through their roots into oxygen and sugars. Plants consume the sugars to build their bodies, and they release oxygen through tiny holes in their leaves. This is why we value plants, and trees in particular: they produce the oxygen that animals and humans need to breathe. This conversion process, photosynthesis, happens on land and in the oceans, where tiny floating plants called phytoplankton perform the same process and produce oxygen that passes into the atmosphere and into the water.

Biologists have tried to work out how much food a large dinosaur would eat each day. They spoke to zookeepers to find out what an elephant eats, which was about 90 lb (40 kg) of plant food a day. If humans eat 2,500 calories per day, that is less than 1 lb (500 g) of food (= 3,500 calories), so the elephant is eating 315,000 calories per day, nearly a hundred times as much as a human eats.

Now, if an elephant weighs 5 tons, and a sauropod dinosaur such as *Diplodocus* weighs 10 tons, did it eat twice as much? The calculations suggest that *Diplodocus* would have eaten about the same amount as a modern elephant, or even less, maybe 66 lb (30 kg) of ferns or horsetails. The reason the dinosaur could eat less was that it had a more efficient system of endothermy (see page 54) and more efficient breathing (see page 64).

DINOSAUR FOOD PYRAMID

Food (energy) passes through an ecosystem usually in the form of a pyramid. There are most primary producers like plants at the base, and these are the food for herbivores.

The carnivores are in smallest numbers because they require a large amount of prey animals (herbivores) for food.

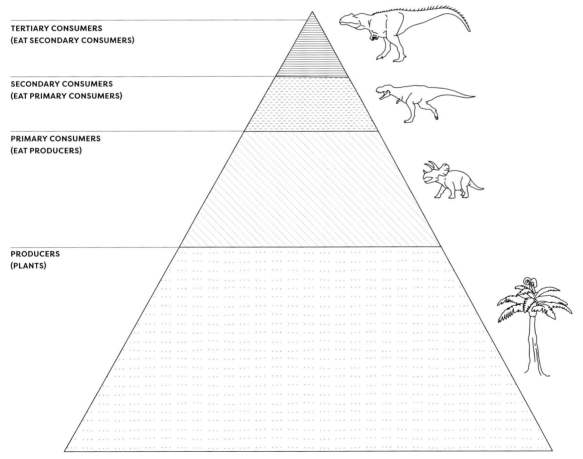

TERTIARY CONSUMERS
(EAT SECONDARY CONSUMERS)

SECONDARY CONSUMERS
(EAT PRIMARY CONSUMERS)

PRIMARY CONSUMERS
(EAT PRODUCERS)

PRODUCERS
(PLANTS)

EACH LEVEL CONSUMES 90% OF THE LEVEL BELOW IT

FORENSICS:
MAKING A DINOSAUR FOOD WEB

It might seem fanciful to try to understand how a dinosaurian ecosystem operated, but that is just what paleontologists want to do. First, we know the different species that occur together from their fossils. So, in the food web shown here, from the Hell Creek Formation of Montana, the key dinosaurs are *T. rex*, *Pachycephalosaurus*, *Triceratops*, *Ankylosaurus*, and a small raptor such as *Dakotaraptor*. Fossils of all of these have been found sometimes lying close to one another in the same layers of rock, so we are pretty sure they all lived side-by-side in their Late Cretaceous world 67 million years ago.

To start the food web, we draw arrows pointing from food to feeder. We know *T. rex* was the top predator and would eat all the other dinosaurs, so all arrows lead to *T. rex*. There is direct evidence in this case: for example, some *Triceratops* bones show teeth marks made by *T. rex*, and there is a famous specimen of a *T. rex* poop, about 3 ft (1 m) long and full of crushed *Triceratops* bones. That's direct evidence!

Then there are other animals including small mammals, lizards, pond turtles, fish, and insects. There are even many other dinosaurs of hugely differing sizes. We don't show all of

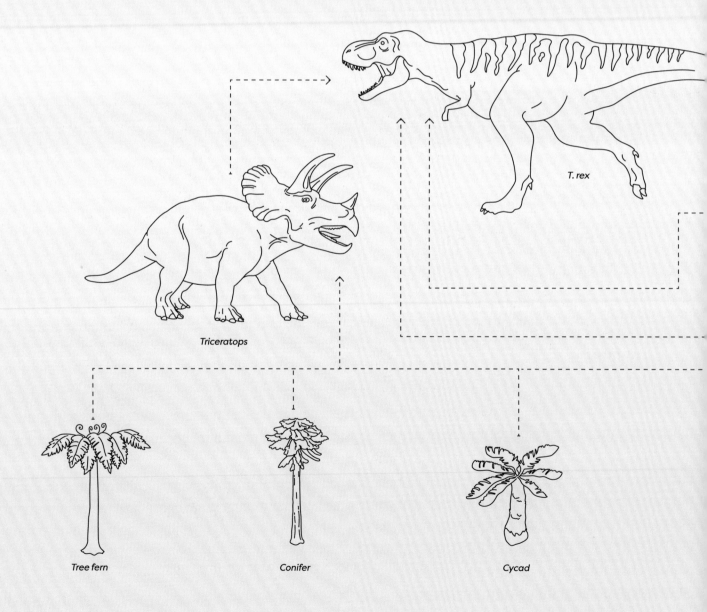

these, to keep the diagram simple. The raptor would have eaten these smaller critters, and while we don't have any raptor poop to be sure, it seems a reasonable assumption because the raptor is clearly a carnivore, with its grasping claws and long, sharp teeth edged with serrations.

Then, working down the food pyramid, there are fossils of all the Hell Creek plants, including seed ferns, conifers, cycads, ferns, and flowering plants. We know from their teeth that *Triceratops* and *Ankylosaurus* were herbivores, specializing in tough stems. The hadrosaurs such as *Edmontosaurus* specialized in conifers, as we know from fossilized stomach contents. Maybe *Pachycephalosaurus* ate some smaller animals as well as plants, so it was an omnivore, or an "eat everything" dinosaur.

NEXT PAGE
On a rainy day near the end of the Cretaceous, this scene is based on fossils form the Hell Creek Formation of Montana. At the front, two brightly feathered predators, *Acheroraptors*, feed on a dead *Pachycephalosaurus*. At the back right, a classic fight scene is shaping up as the horned ceratopsian *Triceratops* squares up to the top predator, *T. rex*.

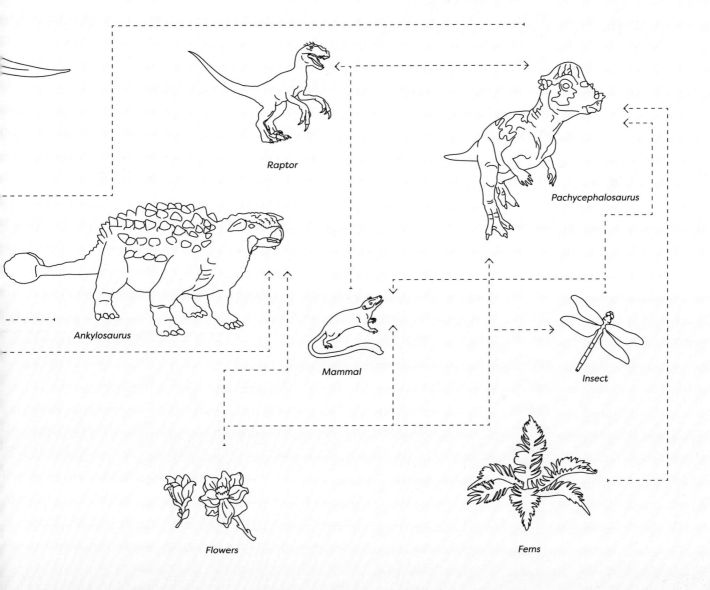

Raptor

Pachycephalosaurus

Ankylosaurus

Mammal

Insect

Flowers

Ferns

PHYSIOLOGY

3
LOCOMOTION

POSTURE AND GAIT

Dinosaurs stood high on their limbs like modern birds and mammals, not sprawling like living reptiles. They could take long strides and keep moving for long time spans.

The ancestors of dinosaurs, and of all reptiles, were sprawlers, meaning that these creatures held their arms and legs partly sideways. We can see this when we look at a modern lizard: when it runs, it bends its body from side to side, and the arms and legs twirl in a crazy way out to the sides. The upper bones of the arm (humerus) and leg (femur) are more or less horizontal and swing back and forth, and the lower parts of the limbs, including hand and foot, have to turn sharply as the animal takes a stride.

In mammals and birds, on the other hand, the arms and legs are directly beneath the body, and the limb movement is simple. The knee, ankle, elbow, and wrist joints are more or less simple hinges, whereas in sprawlers, those joints have to allow more complex swings. Being a sprawler when you are small is not such a bad way to live: a lizard can run fast for a few seconds and shoot under a rock if it is threatened. But this is not such a great adaptation in a larger animal.

SPRAWLING AND PARASAGITTAL LOCOMOTION

Sprawlers cannot breathe and walk at the same time; instead the air is pushed from side to side as the lizard walks.

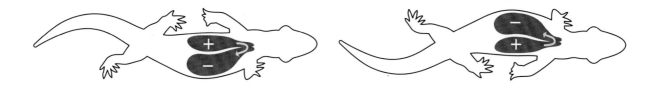

OSTRICH COMPARISON
By comparing the "ostrich dinosaur" *Struthiomimus* (left) with the modern ostrich (right) we can see similarities in its general behavior and running speed.

PARASAGITTAL POSTURE

The upright, or erect, posture of dinosaurs, birds, and mammals is called the parasagittal posture. This means the limbs are directly beneath the body and move parallel to the midline of the animal (its sagittal line). There are two main advantages. First, the animal can take a longer stride because the whole limb is used, not just the lower part. Second, the animal can run and breathe at the same time and so has much greater endurance. A sprawler can only run or breathe, not both; when it moves, the body swings in a major way from side to side, and this forces air from lung to lung and disturbs the breathing. After a few dozen steps, the sprawler has to stop to puff a bit, whereas a healthy human, horse, or bird can run continuously, breathing all the time.

Dinosaurs were parasagittal from the start, and this was probably a major reason for their success. As we saw earlier (see page 37), there was a huge revolution in tetrapod postures in the Triassic period, when nearly all medium-sized beasts switched from sprawling to parasagittal locomotion, which marked a speeding up of life on land.

A parasagittal animal can breathe and run or walk at the same time. Air is pumped in and out of the lungs during the dog's strides.

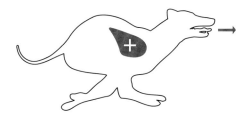
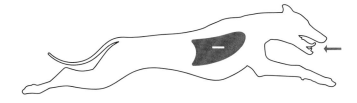

LOCOMOTION

FORENSICS:
MODELING LOCOMOTION

In a well-preserved dinosaur skeleton, the limb bones can tell us a great deal about locomotion. First, the paleontologist fits the bone joints to test the directions and limits of movement. For example, the knee joint, as in humans, has a particular shape, so the bones can move only back and forth, not sideways or allow the knee to bend back.

Then the muscles can be reconstructed, because vertebrates all share much the same leg muscles. So, by comparing dinosaur bones to a close modern relative—birds or crocodiles—the paleontologist knows which muscles were there and what they did. Some muscles pull the leg back (the power stroke in walking), others pull it forward or turn the foot from left to right.

Often there are clear markings on the fossil bone to show where the muscles attached. These markings indicate the size of the muscle and its power—wide muscles that attach over a broad area on the bone provide greater forces than skinny muscles. Muscles only pull by shortening the multiple fibers (they cannot push), so the direction and width of a muscle allows the analyst to work out its approximate effects.

Until a few years ago, that was as far as the paleontologist could go. But now there are dynamic mechanical models that can operate even in a regular desktop computer. In a recent study, researchers from the Royal Veterinary College in London, UK, explored locomotion in the early flesh-eating dinosaur *Coelophysis*.

SIMULATED MOVEMENT
Computer simulations of running locomotion in a modern tinamou bird (*Eudromia elegans*, brown) and extinct theropod dinosaur (*Coelophysis bauri*, green).

20 in (50 cm)

20 in (50 cm)

Using a 3D model of a complete skeleton, they experimented with five thousand random limb postures to tune the models to their most likely solutions.

Then, with the hindlegs pacing along, they worked out what the tail was doing. As a biped moves, the body has to swing from side to side as the whole weight of the body is over the left leg, then the right, then the left, and so on. In a long-tailed animal, the tail switches from side to side to help keep balance. But researchers discovered that the tail swings also helped *Coelophysis* run forward more efficiently. This was compared to the way humans swing their arms as they walk or run; similarly, a dancer or ice-skater may swing their arms to provide additional angular momentum to perform a pirouette or twirl. As *Coelophysis* switched its tail to the left, as the right foot went down in a stride, that swing helped it to move forward in a more stable and energy-efficient manner.

Does this kind of analysis work? There are two reasons it can: first, the researchers used the modern tinamou, a running bird, as a comparative model, and their calculations produced results that closely matched what it does in life. Also, they used standard engineering software that is used in designing machines, robots, and limb replacements, and is well known to produce precise results for those systems.

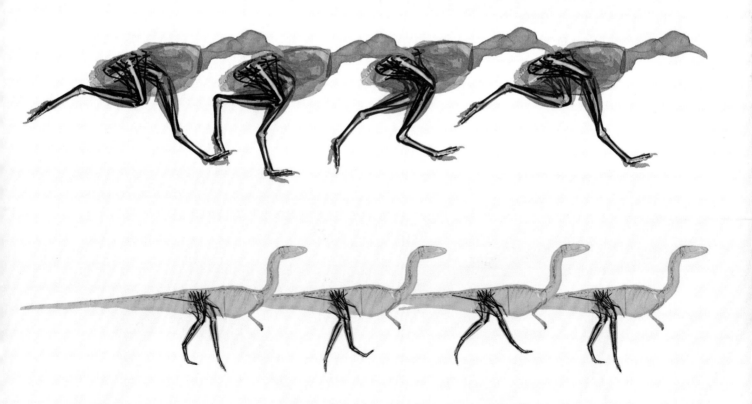

TWO LEGS OR FOUR?

The first dinosaurs were bipeds, using their arms to gather food or fight, but larger forms emerged in the Late Triassic period, and they became quadrupedal to support their heavy bodies.

As humans are bipeds, we may think this is the more advanced kind of posture. Indeed, the first reptiles were all quadrupeds and they used a sprawling posture, but the ancestors of dinosaurs were bipeds from the start, and also quite small, generally less than 3 ft (1 m) long. Being bipedal has its advantages. In humans, it allows us to stand tall and see far. The same may be true for early dinosaurs, but bipedalism also means the arms can be used for other things.

Big individuals weighed 6 tons, more than most adult elephants, and yet they still walked on their hind legs. But famously *T. rex* had silly little arms, certainly not long enough even to reach its mouth, so it's not clear what functions they had.

Among the plant-eaters, as they became larger, several groups such as sauropods, stegosaurs, ornithopods, and ceratopsians went down on all fours. Their arms became stockier and sometimes longer.

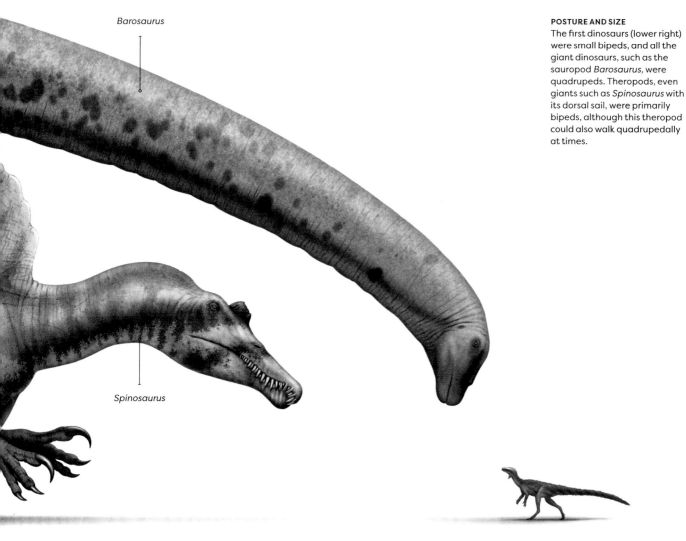

Barosaurus

Spinosaurus

Eoraptor

POSTURE AND SIZE
The first dinosaurs (lower right) were small bipeds, and all the giant dinosaurs, such as the sauropod *Barosaurus*, were quadrupeds. Theropods, even giants such as *Spinosaurus* with its dorsal sail, were primarily bipeds, although this theropod could also walk quadrupedally at times.

LOCOMOTION

TRACKWAYS

Dinosaur footprints and tracks give us unique insights into behavior because they can capture details of walking, and even tell stories of herds on the move or predator attacks.

A 20-ft (6-m) long predator *Australovenator* advances on a herd of 150 chicken-sized coelurosaurs that are drinking from a lake some 100 million years ago in what is now Queensland, Australia. Suddenly, a couple of the small coelurosaurs spot the huge predator and give a chittering warning cry, and all the hundreds of small animals flee in panic, some running toward the looming predator, others tearing off in the other direction.

How do we know all this? It's preserved in the famous "dinosaur stampede" site called Lark Quarry, open to tourists as a covered natural site. The single surface of Early Cretaceous sandstone at Lark Quarry shows some 3,300 individual dinosaur footprints, allowing paleontologists to follow the movements of a single large dinosaur with 20-inch (50-cm) long feet, and hundreds of little critters with feet a tenth of that size.

TELLING FOOTPRINTS
The large plant-eating ornithopod *Muttaburrasaurus* comes upon a herd of small theropods, and they scatter. This scene is based on a fantastic site at Lark Quarry in Australia, where hundreds of footprints tell the story.

ABOVE
Lark Quarry in Australia is such a classic dinosaur footprint site that it's been turned into a museum. A smart new building (above) covers hundreds of footprints, and visitors can get up close to the remarkable fossils inside (right).

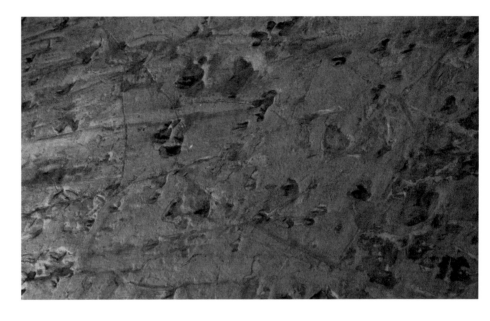

RIGHT
Footprints from small flesh-eating theropod dinosaurs in close-up from the huge trackway surface at the Lark Quarry site. The tiny three-toed footprints show that the dinosaurs were walking from right to left as the photograph is viewed.

This story was challenged in 2011 by Anthony Romillo of The University of Queensland: he identified the "giant predator" as the plant-eater *Muttaburrasaurus*, because the three-toed feet have blunt toes, suggesting hooves, not pointed claws. So, maybe the little dinosaurs ran at another time, or maybe *Muttaburrasaurus* spooked them. Either way, we have thousands of dinosaur tracks all showing amazing detail of some ancient action, and even impressions of skin on the soles of their feet.

Some of the first dinosaur fossils to be illustrated were fossil tracks from the Late Triassic of the Connecticut Valley. The first specimens had been noted around 1800, and Edward Hitchcock (1793–1864), professor at Amherst College, Massachusetts, made a lifelong study, publishing several papers and books from the 1830s onward. He thought they came from giant birds because they had three toes, just like most birds today.

Now, we know theropods and ornithopods were bipeds that made three-toed prints. The massive quadrupeds, such as ankylosaurs, stegosaurs, ceratopsians, and sauropods, made more circular prints, often with short stumpy toes and fingers.

Trackway sites from the American Midwest show that great herds of these animals moved north and south along the shores of the Western Interior Seaway. One sauropod track site is said to show the juveniles were keeping to the safety of the middle of the moving herd, while the large adults trekked along on the outside to protect the young.

TRAMPING ALONG THE SEAWAY

As we saw earlier (see page 16), North America in the Late Cretaceous was divided by a great continental sea that ran from the Caribbean, up through Texas, across Montana and Alberta, and through to the Arctic.

Dinosaurs used to trek north and south along the shores of this seaway, on the "dinosaur freeway," named by University of Colorado dinosaur footprint expert Martin Lockley. He had studied many dinosaur mega-track sites with thousands of footprints, mostly forming identifiable trackways, and he noticed that in many cases, they all ran north to south, not east to west.

At one site near Moab, Utah, Lockley counted over a thousand tracks made by an *Iguanodon*-like ornithopod, over a hundred made by two types of theropods, a hundred by ankylosaurs, as well as tracks from crocodiles, turtles, and pterosaurs. These are trackways, meaning multiple series of prints showing a continuous walk pattern by an individual, so a thousand trackways means more than 10,000 individual prints.

Lockley identified 130 separate track sites in a single rock unit, the Dakota

LEFT
Here, at a site in Denali Natural Park in Alaska, USA, dozens of trackways made by hadrosaurs run in parallel, suggesting a whole herd passed this way, perhaps migrating north or south with the seasons along the shores of the Western Interior Seaway.

BELOW
On the move: A small herd of *Alamosaurus* walk along the edge of the seaway, leaving trails of their footprints. They migrated long distances just as elephants and caribou do today, in search of food. Some footprint evidence suggests the adults guarded the babies, which kept in the center of the moving herd.

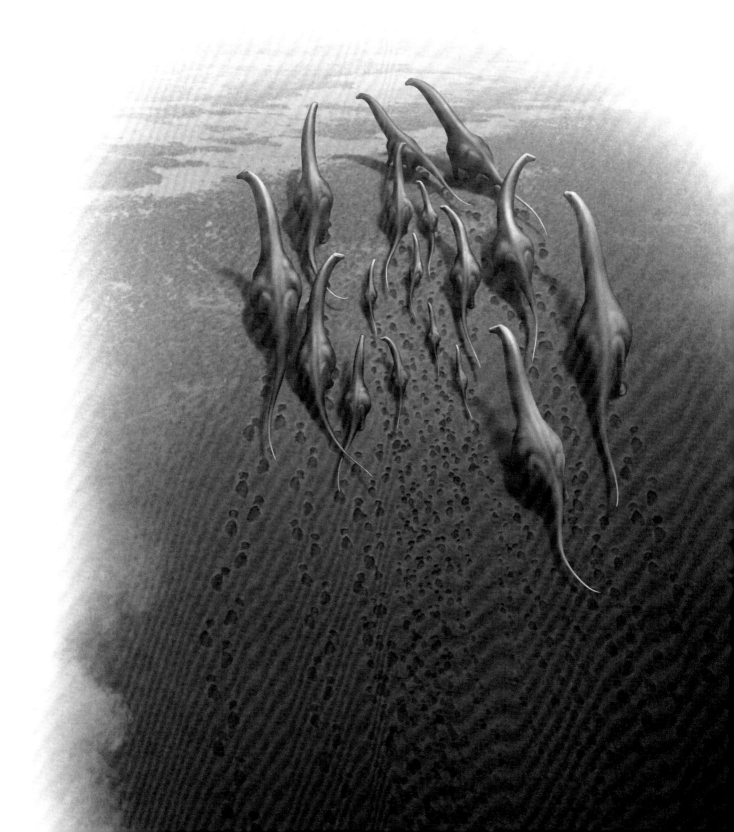

Sandstone, which extends over wide areas of Utah and Colorado. The smaller animals, such as turtles, crocodiles, and many of the dinosaurs, were living at the sites, feeding, drinking, and wandering about. But some of the tracks by dinosaurs, including fifty *Iguanodon* trackways at a site in New Mexico, are all heading northwest, suggesting they were marching along, perhaps part of a long-distance migration.

As the seasons went by, with warmer summers and cooler winters, the supply of plant food would have varied. Remarkably, paleontologists have found similar species of ornithopods all the way from Mexico to Alaska in the Late Cretaceous, and it seems they used the lowlands along the Western Interior Seaway as a migration route of 4,500 miles (7,000 km) in all. Perhaps no single animal walked that far, but the Alaska ornithopods were likely up there only in summer when there was enough food, and they would have headed hundreds or thousands of miles south to present-day Alberta or Montana during the cooler spring or fall.

KEEPING UP WITH MOM AND DAD

Dinosaur trackers can identify roughly which dinosaur made each track, then they can estimate body size from the size of the prints. Many examples show that dinosaur herds were mixed, consisting of adults, juveniles, and babies. Just like large groups of elephant today, these great herds moved from place to place, perhaps in search of food and water, and with the babies and juveniles running to keep up.

Did the parents protect their young? There is wide evidence for parental care in some dinosaurs (see page 176), but what can tracks tell us? In 1941, Roland T. Bird of the American Museum of Natural History published a plan of some remarkable sauropod tracks he had discovered on the Davenport Ranch in Texas. He noticed that all the tracks had been made by a single species of sauropod, and they were all heading in the same direction. More than that, there were some tracks of babies near the middle of the herd.

In 1968, Robert T. Bakker suggested that not only were these dinosaurs moving as a herd, but also that the young ones either knew to stay in the middle of the herd or their parents were fencing them in and trying to keep them safe.

However, this interpretation has not always been accepted. Paleontologists remind us that, although the tracks are remarkable and show a lot of detail about dinosaur locomotion, we can't be sure all these trackways were produced on the same day. It could be that a dozen dinosaurs passed the site over the course of a week or more, each of them following a similar route. On the other hand, the tracks do seem to keep mainly separate from one another, which is more in line with an assumption that they were all walking along together at the same time.

FORENSICS:
TIPPY-TOES TAKE-OFF

Debra Mickelsen, a graduate student at University of Colorado Boulder, found a remarkable series of footprints in the Late Jurassic of Wyoming, which showed a theropod dinosaur walking into a river and shifting from walking to swimming. The track starts with deep prints made as the theropod walked or waded through the shallow waters. Then it wanders into deeper water and the prints become a little less deep, until there are just some light scratch marks made by the three toes. Following these last prints, there are thin scratches made by the claws, then just the toe tips, and finally no trace, as the animal must have swum away.

CALCULATING RUNNING SPEED

How did dinosaurs walk and run? Applying the principles of biomechanics to these ancient animals allows us to understand a great deal of detail about their stance and running.

It's well known that the cheetah is the fastest running animal today. For short distances, it can sprint at 70 miles (112 km) per hour. This is an astonishing speed for any animal to achieve, and three times as fast as the fastest human sprinter. Yet elephants can reach 20 miles (32 km) per hour, not much slower than the human sprinter, but we would see their fastest speed as a lumbering trot rather than a gallop.

So, there's something about body size and speed. There's also something about stride length and speed. If you walk along a beach, you can look back and see your footprints. If you walk slowly, the footprints are close together. If you speed up, your stride length becomes longer, and at full speed, your footprints are more spaced out.

In the 1980s, Robert McNeil Alexander (1934–2016), an expert in biomechanics at

TOP SPEED OF LAND ANIMALS

The small dinosaur runs faster than the large dinosaur. All the evidence suggests that the giants like *T. rex* here could not sprint and so plodded along.

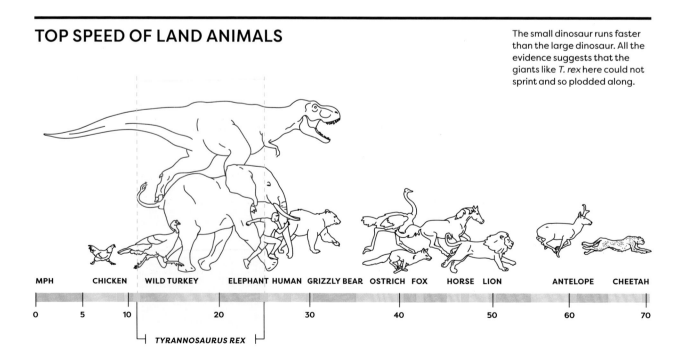

DINOSAUR BEHAVIOR

Leeds University, UK, realized that it should be possible to work out a simple formula to calculate speed from stride length, and, if the formula worked for modern animals, it would also work for extinct animals such as dinosaurs. In the end, he came up with this formula:

$$u = 0.25 g^{-0.25} = 0.5\, d^{1.67}\, h^{-1.17}$$

where u is velocity, g is gravity, d is stride length, and h is hip height.

The first concern McNeil Alexander had was whether his formula worked. He got together all the animals he could find and, with his family, they went to the beach. He, his wife, and his children all walked, ran, and dawdled across the beach, and he measured their speed with a stopwatch. Then they tried the family dog, a horse, a cat—the formula worked every time, whether the animal was small or large, ran on four legs or two. Therefore, it could be applied to dinosaurs. We know gravity but would have to measure hip height and stride length. Hip height comes from the animal that made the tracks, which is usually easy to estimate. Stride length is simply measured from the spacing of footprints in the trackway.

The Alexander formula was revolutionary. Paleontologists could apply it to all the fossil trackways they knew, and they could at last answer some long-running debates. For example, could *T. rex* run as fast as a racehorse (45 miles/64 km per hour) or did it slouch along at a much slower speed? Typical calculated speeds for dinosaurs range from 3–9 miles (5–15 km) per hour for walking dinosaurs, about equivalent to human walking and trotting speeds. Sometimes, higher speeds of 22–35 miles (36–54 km) per hour were calculated for smaller, flesh-eating dinosaurs that were in a hurry to catch their lunch.

What about *T. rex*? All the trackways indicate a maximum speed of 17 miles (27 km) per hour, faster than a wild turkey, about in line with an elephant and a regular human (not an athlete) running to catch the bus.

FORENSICS:
DINOSAUR WALKING KINEMATICS

Tracks have complicated relationships to the mud or sand the dinosaurs were walking on. If the surface of the ground was dry and hard, the dinosaurs would leave little impression, but the softer and gloopier the sediment, the deeper their feet would sink in. The tracks can then be seen on many different layers even a foot (30 cm) below the surface. These are called undertracks.

Stephen Gatesy of the University of Rhode Island was able to explore the internal structures beneath a dinosaur track to see the combined effects of gloopy mud with the multiple impressions seen in the undertracks.

Scanning and reconstructing all the layers beneath tracks from Connecticut, and similar dinosaur tracks from the Triassic of Greenland, Gatesy and collaborators were able to animate the whole process. They could see how the dinosaur foot had gone in then sunk through layers of mud. As the animal moved forward, it pulled its foot out of the mud, and the mud flowed back around the withdrawal mark, leaving a trail produced by the long middle toe.

Computer animations then showed how the three-toed bony foot must have worked as the animal moved forward. As the animal put one foot forward and down, the three slender toes spread out as far as they could. On a firm surface, the foot would not sink, but when it was muddy, the wide-spread toes did not prevent sinking. This sinking and pulling while the animal ran took a lot of energy, as you know if you walk on mud or soft sand. As the foot sank into the mud, the animal pulled its toes together, then curled the foot back in a kind of fist to withdraw it from the mud with least resistance.

Gatesy and collaborators now use a variety of laboratory equipment to

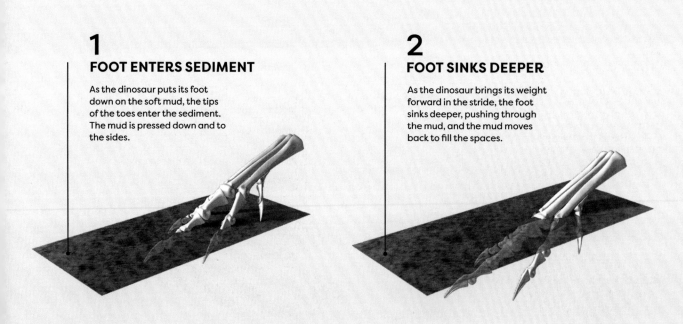

1
FOOT ENTERS SEDIMENT

As the dinosaur puts its foot down on the soft mud, the tips of the toes enter the sediment. The mud is pressed down and to the sides.

2
FOOT SINKS DEEPER

As the dinosaur brings its weight forward in the stride, the foot sinks deeper, pushing through the mud, and the mud moves back to fill the spaces.

understand the kinematics, meaning the relative motions of structures, of dinosaur feet in walking and running. They can X-ray the fossil tracks to see the undertracks and how they change shape. They do experiments with modern running birds such as emus or chickens, to see how they change the attitude of the foot in running on different kinds of sediment. They can even use X-ray movie cameras to view the bird's feet as they penetrate into the sediment below.

The researchers also carry out experiments with model dinosaur feet made from wire and sediments of different kinds (varying the size of grains and the amount of water) and recording the exact three-dimensional shapes of the resulting tracks. This kind of forensic paleontology enables researchers to understand how dinosaurs moved and how they coped with different ground conditions.

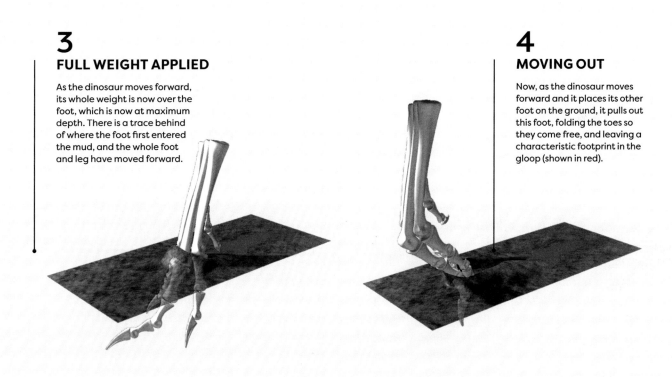

3
FULL WEIGHT APPLIED

As the dinosaur moves forward, its whole weight is now over the foot, which is now at maximum depth. There is a trace behind of where the foot first entered the mud, and the whole foot and leg have moved forward.

4
MOVING OUT

Now, as the dinosaur moves forward and it places its other foot on the ground, it pulls out this foot, folding the toes so they come free, and leaving a characteristic footprint in the gloop (shown in red).

GIANT SIZE AND SUPPORT

Dinosaurs are famous for their size. Close study of their skeletons shows how they combined massive bones with efficiencies in metabolism to be able to achieve such huge sizes.

The first dinosaurs were relatively small—some of them little more than 3 ft (1 m) long—lightweight predators that were bipedal. Indeed, bipedality was the primitive state (see page 81), and later dinosaurs became quadrupedal generally when they became larger. Amazingly, the theropod giants such as *T. rex* and *Spinosaurus* were bipeds, and they never became quadrupeds, even though they weighed 5 tons or more.

But the real giants were the sauropods. These started, like all dinosaurs, as small bipeds, and some of the Late Triassic forms such as *Plateosaurus* became large rather rapidly. Sauropods, which were quadrupedal all the time, also arose in the Late Triassic, then diversified in the Jurassic, some of them reaching huge sizes by the Late Jurassic.

The Morrison Formation in North America and the Tendaguru locality in Tanzania provide examples of some of the best-known sauropods from the height of their success. Beasts such as *Apatosaurus, Brachiosaurus, Brontosaurus, Camarasaurus, Diplodocus,* and others would all have strained the weighing scales to the limit. *Diplodocus* was over 80 ft (24 m) long. *Brachiosaurus* and *Giraffatitan* were 76 ft (23 m) long and their heads could tower 39 ft (12 m) above the ground. Current estimates indicate they weighed anything from 10–100 tons.

Every aspect of the skeleton of *Brachiosaurus* shows its adaptation to huge size. In particular, the limbs are pillar-like, located straight under the body for maximum weight support, just as in an elephant. In addition, the hands and feet are much shortened and simplified, and there would have been a thickened pad beneath these. So, instead of the foot being long and slender, as in other dinosaurs, fingers and toes were shortened, bones of the wrist and ankle reduced, and the hand and foot were not much wider than the whole limb. In most sauropods, the first finger and the first two toes bore long claws that might have been used in digging. The other fingers and toes do not have claws, but end in small hooves, presumably for protection as they walked.

The bones of the backbone were very large, forming a strong bridge-like structure between the fore limbs and hind limbs. In addition, as we saw earlier (page 65), the vertebrae and rib heads in sauropods were pneumatic, containing air cavities. This was connected with their birdlike respiration system, but also reduced the weight of the vertebrae to at least half of what it would have been if the bone were solid.

ESTIMATING THE WEIGHT OF A SAUROPOD

How much did these monsters weigh? Paleontologists have tussled for years over estimates of body weight, or mass. It's difficult. For weight, do you measure the length and scale it up somehow by comparing with a modern elephant or crocodile?

Ned Colbert (1905–2001) from the American Museum of Natural History had a great idea when he realized he could use the Archimedes Principle. This is the observation

SAUROPOD SKELETONS

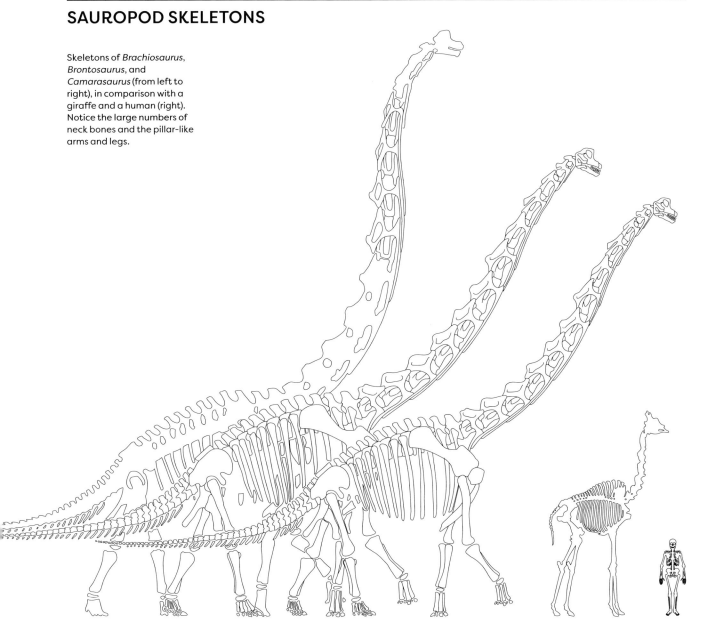

Skeletons of *Brachiosaurus*, *Brontosaurus*, and *Camarasaurus* (from left to right), in comparison with a giraffe and a human (right). Notice the large numbers of neck bones and the pillar-like arms and legs.

that volume and mass are related. It is said that the ancient Greek philosopher Archimedes was worrying over this question while having a deep bath, when he realized that his body volume was equivalent to his weight. As he sank into the full bath, the volume of water equivalent to the volume of his body spilled over the edge. It's said that he leaped from the bath and ran naked through the town shouting, "Eureka! I have found it out!"

Colbert took accurate plastic models of dinosaurs and dangled them in a container with a liter (33 fl oz) of water. He measured how much was displaced to get the volume. Knowing that 1 liter of water is roughly equivalent to 1 kg (2 lb) of mass, he had the weight of the model. He then scaled this up to the original size of the animal. (There's an assumption here, which is that dinosaur flesh has the same density as water, but in fact that's more or less true.)

Colbert found masses of: 2 tons for *Allosaurus*; 7 tons for *Tyrannosaurus*; 12 tons for *Diplodocus*; 30–35 tons for *Brontosaurus*; and 85 tons for *Brachiosaurus*. Since 1962, when Colbert published these estimates, the work has been repeated many times. One dispute was over whether dinosaurs were fat or thin—if they were fat they might have weighed twice as much as if they were skinny and their ribs stuck out. Also, their mass is not evenly distributed as solid flesh in the way Colbert thought—there are big air spaces in the mouths and lungs, and some areas, such as the hips, have lots of dense bone. Most of Colbert's estimates are considered reasonable today, but 85 tons for *Brachiosaurus* was maybe a bit high, and modern estimates range from 28–58 tons.

Instead of plunging plastic models into the bath, paleontologists have realized they can use a proxy for body mass to obtain

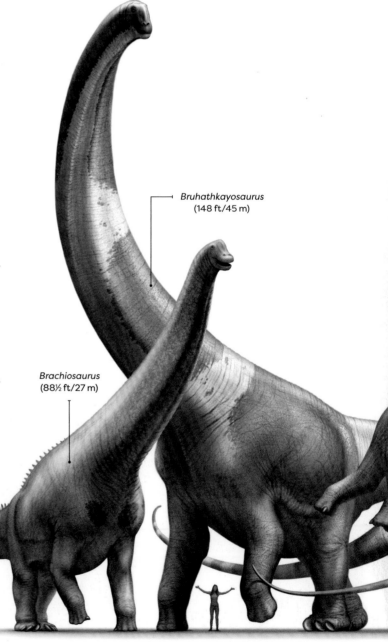

Bruhathkayosaurus (148 ft/45 m)

Brachiosaurus (88½ ft/27 m)

SAUROPOD PARTY TIME! Here, a group of sauropods from different parts of the world and different times in the Jurassic and Cretaceous show the amazing variety of sizes and shapes. The lengths, from snout to tail tip, are given in brackets below the names of each animal.

accurate estimates. A proxy is a stand-in, and the easiest to measure is from the size of the main bones in the arm and leg, the humerus and femur. The circumference, or distance around, the head of the humerus and femur is proportional to body mass because, in life, these bones have to support the huge body weight. By plotting this measurement against body masses for living mammals and reptiles, researchers found a regular relationship between bone circumference and body mass; then dinosaur measurements could be plotted, and their body mass estimates read. In the graph, the sauropods have bone circumferences of 4–79 inches (10–200 cm), equivalent to body masses of 1–100 tons.

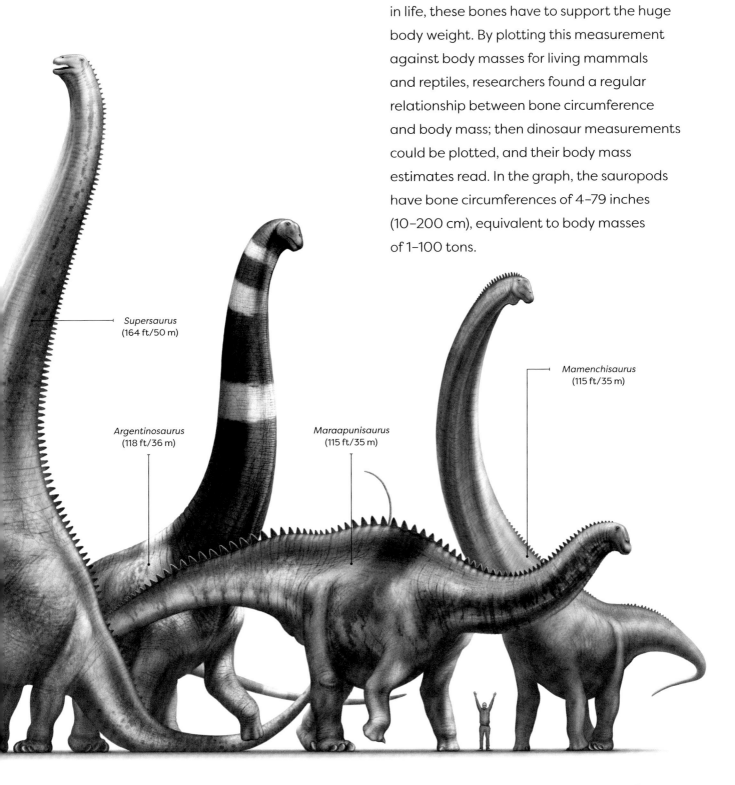

Supersaurus (164 ft/50 m)

Argentinosaurus (118 ft/36 m)

Maraapunisaurus (115 ft/35 m)

Mamenchisaurus (115 ft/35 m)

LOCOMOTION

ON BEING A GIANT

How could a sauropod weighing 50 tons survive? After all, we know that modern elephants have to spend most of their time looking for food, and they may walk hundreds of miles to follow the food supply. German paleontologist Martin Sander and his team have worked out that sauropods survived by having a smart way of saving energy, through a combination of their long necks, small heads, birdlike respiration, high basal metabolic rate, and egg laying. Let's look at each of these in turn.

The long neck allowed the sauropod to browse plants over a wide area, up and down and left to right, without budging. Standing still for hours on end and moving the neck and head only while they fed was a smart way to conserve energy.

The small head was a consequence of having a long neck and the difficulty of supporting a large head. Sauropods swallowed their food whole, so they snipped leaves and stems with their teeth and gulped it all down without chewing or any other processing. Sauropods had big rib cages, so probably had huge guts, and their food would have rumbled and sloshed all day as it was digested, and they maybe produced great flops of semiliquid poop. Or, perhaps like modern crocodiles and birds, they conserved water and made solid, hard droppings.

We've already seen that dinosaurs had birdlike respiration (page 64), and this was hugely important for the sauropods.

Sauropods almost certainly had a high metabolic rate, meaning they were endothermic (see page 54). Evidence for this comes from their bone histology (see page 57) and from their growth-rate curves (see page 188), suggesting a typical sauropod reached adult size at an age of fifteen to twenty years, and that they put on 0.5–2 tons of weight per year. This phenomenal rate of growth indicates high metabolic rate, and that they used oxygen and food in an efficient manner.

The final point about sauropods is that they laid eggs, and small eggs at that. Sauropod eggs were about the size of an American football, and hatchlings were about 2 ft (0.5 m) long, compared to parents who were 50–260 ft (15–80 m) long. The mother laid a dozen eggs in a nest scooped out of the soil, and maybe covered them before wandering off. Parental care was probably limited or nonexistent (see page 176), which contrasts with the long-term efforts by mammals. Laying small eggs and abandoning the babies saves a huge amount of energy and risk for the mother. Of course, the baby sauropods were at risk from predators, but if the mother laid between ten and twenty eggs per year, only one or two had to survive for her breeding to be rated as successful.

So, it seems that sauropods, and other giant dinosaurs, owed their huge size and their evolutionary success to a combination of these five special features. Perhaps mammals aren't always top of the evolutionary tree after all!

FORENSICS:
HOW TO BE A GIANT

This flow chart shows the key features that allowed sauropods to be so huge. It is mainly about their gigantothermy (being warm-blooded because they were huge) together with minimal parental care (abandoning their babies). Their huge size allowed them to live on quite low-nutritional value plants, processing these without chewing and without gastroliths, and saving energy in feeding by standing still for much of the time and just swinging the neck around to grab food. Sauropods had to be warm-blooded to enable their fast growth rate so they could reach sexual maturity within 20 to 30 years.

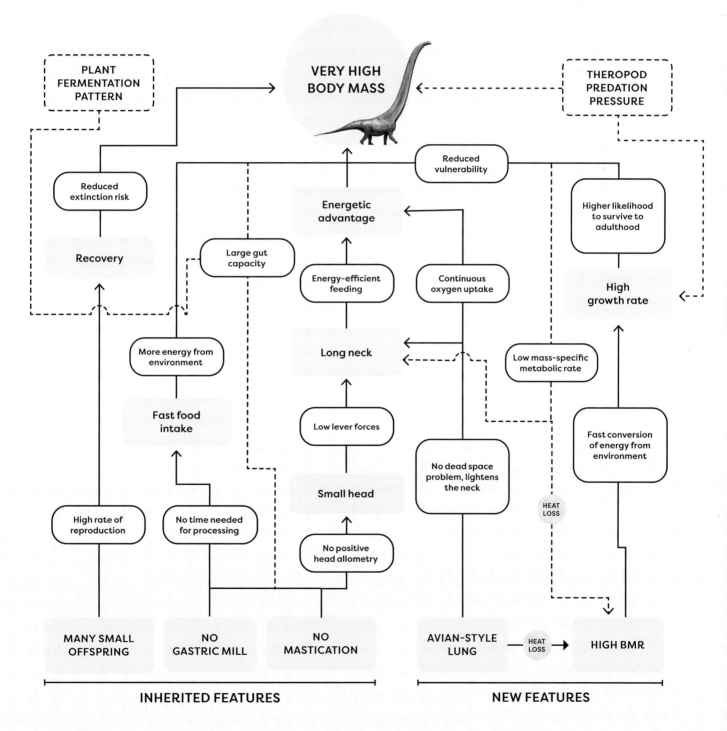

ORIGINS OF FLIGHT

At one time it was thought that only birds, bats, and insects could fly, but new studies show that flight originated several times among dinosaurs.

Birds evolved from dinosaurs, and flight originated at this point. Right or wrong? In fact, wrong: flight originated maybe five times in different dinosaur groups, as shown in a 2020 report by Rui Pei and colleagues, from Beijing. They looked at all the little feathered theropods and birds of the Jurassic and Cretaceous, and showed that more groups had crossed the threshold from gliding to powered flight than anyone had expected.

It's important to understand the use of words here. Biologists normally use the term "flight" to mean any kind of flying, whether by a bird, gliding lizard, or even a twirling plane-tree seed. In common speech we use "flight" to mean "powered flight," where the animal has large wings and specialized flight muscles to beat the wings up and down and stay up in the air for long spans of time.

Today, there are many gliding lizards, snakes, and mammals, all of which live in and among trees and use their expanded membranes to glide. These are great evolutionary adaptations, whether an animal is predator or prey—to extend a leap allows them to escape a predator or snatch a flying insect as they zip by. This is probably what was going on in the Jurassic, as some groups of theropods perhaps returned to the trees in search of insect prey.

In the evolution of birds, the paravian theropods, the dinosaurs closest to birds, showed many kinds of flight. They had powerful hands and arms, and these were lined with pennaceous feathers, just as in modern birds (see page 58). Early experiments on models of dinosaurs such as *Microraptor* showed they certainly could fly (see page 42).

In their 2020 study, Pei and colleagues looked at all the paravians and used a standard ratio in aerodynamics called wing loading, which is body mass divided by wing area. The threshold value is about 10 g per cm^2: above this, the body is too heavy to be kept aloft by the wings; below this, and the animal has the potential to fly. Of course, their test doesn't tell us that *Yi qi*, *Microraptor*, *Rahonavis*, *Anchiornis*, and *Archaeopteryx* did fly, but the fact their wings were large enough to fly suggests they very well might have done. After all, why have big wings if you don't use them?

THE FIRST BIRD
Archaeopteryx spreads its wings as it comes in to land. This dinosaur-bird can fly almost as well as a modern bird, but landing is always risky. The bird has to slow down and stall, meaning it might crash-land. It extends its legs forward and spreads the wings fully, trying to keep control as it approaches its perch.

Are these estimates of wing loading for the fossils accurate? The wing area is known accurately in most cases because the wing bones, membranes, and feathers are well preserved, and so the area can be measured from a fully stretched-out wing fossil. Body mass is an estimate, but Pei and colleagues were careful to use a range of values and to compare their fossil animals to living animals. For example, *Archaeopteryx* is the size of a modern raven and so its body mass estimate is taken to be the same.

PRINCIPLES OF FLIGHT

Birds and aircraft obey the same principles of flight. The wing is an airfoil, meaning it has a particular shape in cross section, with a solid, thick leading edge and a tapering trailing edge. As a bird moves forward, it is subject to four forces: *gravity* pulls it down, which is countered by *lift*. Its forward motion is the *thrust*, but this is slowed by *drag*, all the kinds of friction that can slow the bird down. To be efficient, birds and planes are shaped to minimize drag and maximize thrust, in other words gaining forward movement with

FLYING DINOSAURS
The little four-winged *Microraptor* (left) and *Rahonavis* (right) look very like birds, but they are in fact not birds, and yet evidently could fly.

minimum use of muscle energy or fuel. Also, the balance between gravity and lift has to be just right, or the flying object might fall out of the sky!

The secret of an airfoil is that it provides both lift and, in the case of birds, thrust. An aircraft wing is fixed and thrust comes from the engines. As the bird or plane moves forward, air flows back over and under the wing. The airstream over the top of the wing travels further and faster than below, causing a reduction in pressure over the wing. The wing (and the bird or plane) are lifted up into the lower-pressure area. Maintaining just enough pressure difference keeps the flyer aloft.

Humans have tried to build flapping-wing aircraft, but these were rarely successful, and sadly frequently fatal for the intrepid pilots. But this is what birds do, and they can combine lift and thrust in a single wing.

They have powerful muscles to pull the wing down and up, the main one being the pectoralis muscle, which creates downward movement. As the wing sweeps down and back it pushes the body both up and forward. In the downstroke, the bird spreads its wing as wide as it can to get as much thrust as possible, then it partly folds the wing feathers inward to sweep the wing back up and forward for the next downstroke.

Birds have very different wing shapes depending on their adaptations. Large seabirds, for example, have long narrow wings that are ideal for soaring—flying high on rising air currents and without beating the wings—and good also for flying for days on end over the ocean. On the other hand, forest birds such as owls have short broad wings to allow them to dodge among the trees. Some small birds such as hummingbirds have tiny wings that beat rapidly to allow them to maneuver around plants and insert their beaks into flowers, to feed on nectar. It's likely that flying dinosaurs and early birds did not have such subtle adaptations.

COULD *MICRORAPTOR* FLY?

Most striking of the feathered dinosaurs was *Microraptor*, a close relative of the dromaeosaurid *Deinonychus*. When it was found in the year 2000, paleontologists were intrigued by the fact it had four wings, all liberally lined with beautiful flight feathers. Surely it could fly? Some experimenters assumed it did so with the four wings in a single plane, like a kite, whereas others thought the hind wings might have been held lower and so it would have flown something like a biplane from the First World War.

Initial experiments were uncertain. For example, Colin Palmer at the University of Southampton, UK, planned to test how *Microraptor* flew, and applied his practical engineering knowledge to making tests. He built a life-sized model from foam and stuck feathers from modern birds firmly along the wings. He then tested the model in a wind tunnel in which he could vary the wind speed, orientation of flow, and model posture, allowing the limbs either to dangle down or to stick out sideways. These experiments showed that the best plan for a *Microraptor* was to jump off its perch with its legs sprawled out to the sides, then to let them drop down to a dangling position as it began to glide.

At the time of their study, in 2013, Palmer and colleagues assumed that *Microraptor* was

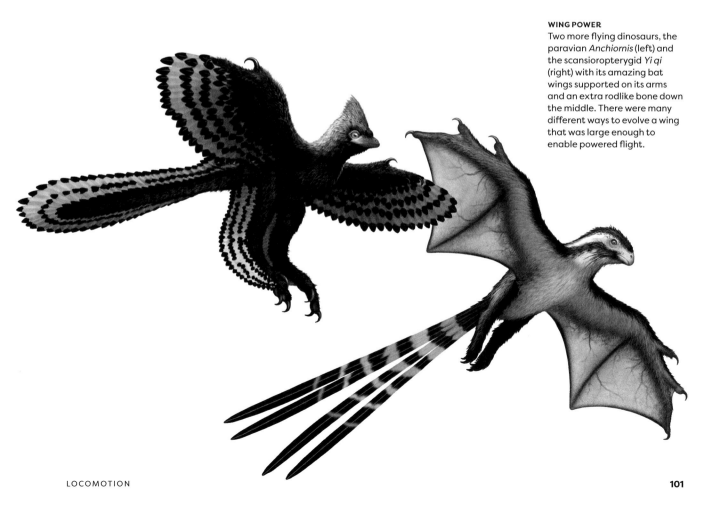

WING POWER
Two more flying dinosaurs, the paravian *Anchiornis* (left) and the scansiopterygid *Yi qi* (right) with its amazing bat wings supported on its arms and an extra rodlike bone down the middle. There were many different ways to evolve a wing that was large enough to enable powered flight.

a glider, not a powered flyer. They worked out in any case that if it was flying around in the treetops, it probably didn't need powered flight, because it could launch and glide over long distances. Like all the small theropods, *Microraptor* presumably had good three-dimensional, or binocular, eyesight as we do, meaning it could judge distance and hit a tree at the end of its flight with accuracy.

The 2020 analysis by Pei and colleagues shows that *Microraptor* could have been capable of powered flight, so it maybe saved energy by gliding for most of the time, but could make small flaps of its wings to adjust its direction or extend the flight just a little further when it had to.

COULD *ARCHAEOPTERYX* FLY?

The answer would seem to be obvious: of course it could! It had wings and feathers and was a bird, so why wouldn't it fly? However, since 1861, when the first specimens were found, scientists have argued the case. In those early days, we didn't have the wealth of fossils of feathered paravians we have now, and some assumed that feathers and wings had perhaps come first, flight later. Others noticed that the breastbone, or sternum, of *Archaeopteryx* was very low.

In modern flying birds the sternum is deep, and it is the main site of attachment of the flight muscles, the powerful pectoralis that causes the downbeat, and the supracoracoideus that causes the upbeat. Anyone who has carved a chicken or turkey knows the sternum and these

BIRDLIKE SKELETONS

At first, the skeleton of a pigeon (far right) looks quite different from the small predatory dinosaur (left), but *Archaeopteryx* is the perfect intermediate. It still has the dinosaur-like teeth, big hands, small bones in the hip area, and a long bony tail. Comparing the first bird, *Archaeopteryx*, with the modern pigeon shows how the hip girdle was modified and fused to reinforce the body for hard landings, the bony tail was shortened, the teeth were lost and replaced by a horny beak, and the hand became reduced to three short fingers, meaning most modern birds cannot use their hands in climbing trees or grabbing things.

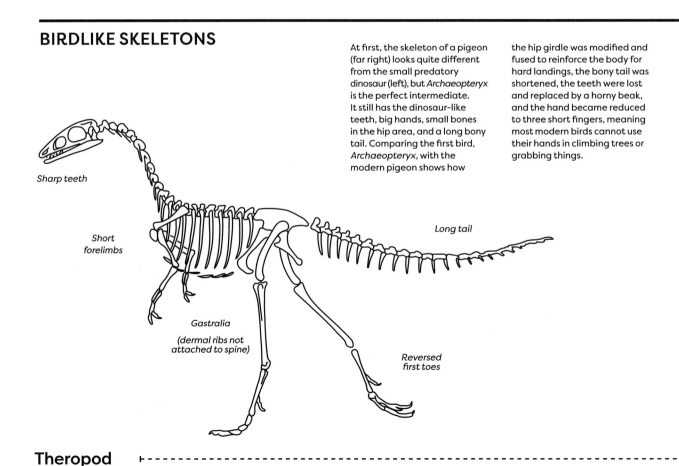

Sharp teeth

Short forelimbs

Gastralia (dermal ribs not attached to spine)

Long tail

Reversed first toes

Theropod

two muscles. The pectoralis makes up most of the breast meat, and stuck down the side is the much smaller supracoracoideus muscle.

Archaeopteryx was a medium-sized bird, about a foot (30 cm) long and standing perhaps 10 inches (25 cm) tall. Its skull was rather dinosaur-like, with small teeth along the jaws, not a beak, but already with large eyes and evidently excellent vision. The wings were powerful and lined with twelve primary and twelve secondary flight feathers, as in many modern birds, but *Archaeopteryx* retained three powerful fingers with claws on its wings, as in dinosaurs. The pelvis and hind limb were also more dinosaurian than avian, lacking all the fusion and strengthening seen in modern birds to enable them to land without breaking their legs.

Without a large sternum, *Archaeopteryx* probably had relatively smaller flight muscles than a modern bird of the same size. Further, with its long bony tail, theropod-like pelvis, and teeth, it had not acquired all the characters seen in modern birds

The Late Jurassic and Early Cretaceous woods in which *Archaeopteryx* and many of the flying feathered paravians lived were probably home to insects, spiders, and other bugs. These little dinosaurs and birds chased and snatched their prey from branches, and may have leaped, glided, and flapped from tree to tree, snatching their prey as they flew off.

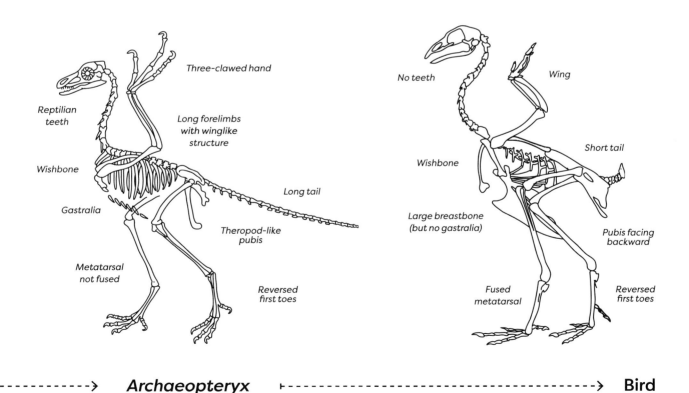

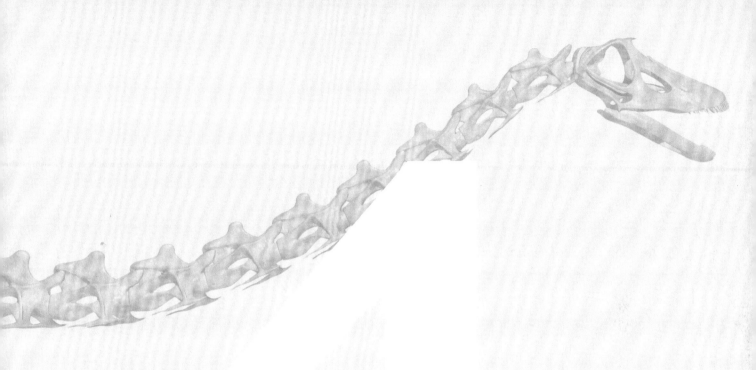

SENSES AND INTELLIGENCE

DINOSAUR BRAINS

Dinosaurs are not famous for their huge brainpower; in fact, most of them had quite small, reptilelike brains that were good for their senses but not for deep thinking.

It's important to realize that dinosaur brains did not fill up their whole skull. In fact, most of the skull was filled with jaw muscles and sensory systems, such as a large nasal cavity for smell, and large eyeballs. In fact, the brain filled a small box in the center near the back, called the braincase. We can think of a typical dinosaur skull as like a shoe box with a matchbox inside, and the matchbox is the braincase. In birds and mammals, on the other hand, the brain has swollen to fill up much of the skull, and the jaw muscles, for example, have been pushed out of the skull to sit on the outside.

Nearly all we know about dinosaur brains comes from internal impressions of the brain, which usually disappears over time as it is made from soft issue. Paleontologists call the impression of the inside of the braincase the endocast. It shows the rough outline of the shape of the brain, which is largely located inside the braincase at the back of the skull, but the olfactory areas, associated with the sense of smell, can extend far forward between the eyes.

Care is still required in interpreting the endocast because it may show more than just the brain. In life, the brain is surrounded by protective tissues called the dura mater, which protect the brain, especially from thumps on the head. In birds, the dura mater is thinner than in crocodilians, probably because bird brains are crammed so tightly into the skull. However, we can't be sure whether dinosaurs had a thick or thin protective dura mater. Some theropods, close to the origin of birds, probably did have birdlike brains.

As we shall see in this chapter, the brain can tell us something about the intelligence and the senses of dinosaurs. There are different areas of the brain in modern animals, with definite functions—for example, for vision, smell, hearing, and balance—and we will explore these further.

Usually there's nothing left of the dinosaur brain, but a discovery reported in 2016 seemed to change that. In a specimen from the Early Cretaceous of southern England, Martin Brasier (1947–2014) of the University of Oxford and colleagues reported what might be a portion of

PRACTICAL BRAINPOWER
The brain of this *Iguanodon* plant-eater is not large. Most of the brain is to control the arms and legs and muscles, and to power its good eyesight. Deep thinking is not part of its repertoire.

the actual brain tissue of the ornithopod dinosaur *Iguanodon*. The whole specimen was just a hand-sized pebble on the beach, but in it they identified part of the skull roof, and below that the dura mater and other tissues around the brain. Underneath these, they identified actual brain tissue and networks of tiny blood capillaries, part of the supply to this ancient dinosaur's brain.

COMPARING DINOSAUR AND BIRD BRAINS

The small theropod dinosaur *Tsaagan* shows a relatively large brain, buried deep within the skull, and with especially developed sensory areas for sight and smell. But the modern bird (below) has an even larger brain, with high-powered senses of sight and hearing, as well as much stronger coordination areas of the brain, which are necessary for the complex life of a flying animal.

Tsaagan
(non-avian dinosaur)

New Caledonian crow

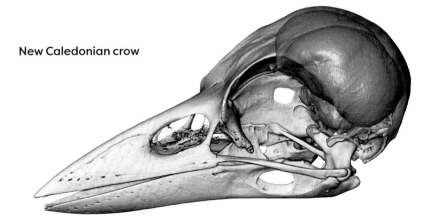

SENSES AND INTELLIGENCE

FORENSICS:
STUDYING ANCIENT BRAINS

In the early days of paleontology, the only way to study a dinosaur brain was to look at the rock inside a fossil skull. After the dinosaur died, sometimes its braincase would fill with sand, and over time this turned into rock and gave a good impression of the shape of the brain. But there's a big risk that you would damage the skull if you tried to remove this brain cast.

The next idea was to clean up a dinosaur skull completely, then block the openings and fill up the braincase with dried peas or lentils. The peas or lentils are tipped out and are then poured into a measuring bottle for an idea of the volume of the brain. However, these methods do not give the full detail.

The rapid spread of computed tomography (CT) methods changed everything. CT scanners were developed first for medical use, so physicians could look at a detailed 3D scan of a sick patient and work out what was wrong. Indeed, brain surgeons use 3D scan models of patients' brains as a means of testing the delicate surgery needed to locate and remove cancers, for example.

In paleontological research, many scans have now been made of dinosaur skulls. It's best to choose a specimen that is in good shape and has not been crushed. Then you can see all the detail of the different internal portions of the head. There are three steps in understanding a dinosaur brain.

First, the paleontologist works on the X-ray CT scans to make a 3D model of the endocast, to get the overall shape of the brain, dura mater, nerves, and other structures. This requires care because the contrast between bone and rock may not be great, and

LEFT
Scanning the braincase of the ceratopsian dinosaur *Triceratops* using a medical whole-body CT scanner. Large fossils require large scanners.

RIGHT
Understanding the brain of the Late Triassic sauropodomorph *Thecodontosaurus*. First the braincase fossil is scanned and a detailed 3D model produced. Then the scans are used to look inside the braincase to see the endocast (cast of the brain), and the endocast is separated. In this image, brain tissue is shown in blue, nerves in yellow, and the semicircular canals, mainly for balance, in pink.

delicate structures such as nerves have to be followed carefully.

Second, the paleontologist tries to interpret the endocast and remove the dura mater to see the brain in more detail. This is a difficult task and can only be done by guessing how the lumps and bumps on the surface of the endocast match original brain structures, and how thick the dura mater was. At least we know from comparison with modern bird and crocodilian brains what to expect. Some dinosaur endocrania show impressions of blood vessels, suggesting the dura was thin at those points, because the blood vessels indicate this was actually the surface of the brain itself.

Third, the paleontologist can then read something about the senses of the dinosaur by looking at relative areas of the brain. It's usually assumed that the relative sizes of different brain areas, such as for smell, sight, and hearing, can actually be measured and that the sizes indicate their importance in life, for example through more focus on smell or vision.

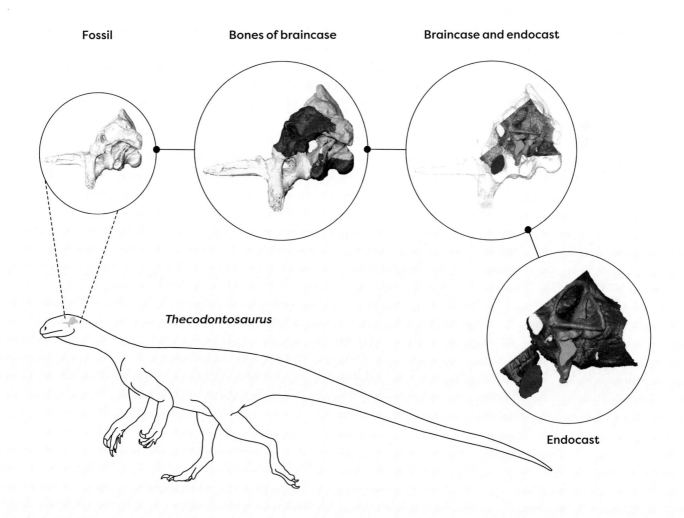

BRAINS OF MODERN REPTILES AND BIRDS

Modern reptiles and birds have brains that look different, but in fact they share many features and so are great models for interpreting dinosaur brains.

If you compare the brains of modern vertebrates, they share a common pattern. Certainly, birds and mammals show great expansion of the forebrain, but all the key parts can be seen. There are four main parts to the brain, and these stand in a row from back to front.

The medulla oblongata is the most basic part of the brain, connecting the spinal cord to the rest of the brain, and is responsible for fundamental body functions such as keeping the heart beating, as well as controlling blood pressure and breathing.

The cerebellum comes next, and it is also used in motor control, particularly helping the body keep its balance, while dealing with basic emotions such as fear and pleasure. The "pons" (Latin for "bridge") connects the brain to the spinal cord and helps regulate basic functions of the body such as breathing and the cycle of sleeping and waking.

The midbrain connects the back and front parts, and has important roles in all the sensory systems, such as vision and hearing, as well as sleep/wake cycles, alertness, and temperature regulation of the body.

The forebrain, or cerebrum, is the site of all the functions we might group together as "intelligence," including reasoning, intellect, memory, language, and personality. This part of the brain is largest in mammals and birds but small in fishes.

Other brain areas control the senses. The optic tectum is the brain area that controls vision, and the olfactory bulb at the very front of the brain is the area in which senses of smell and taste are processed. Humans, for example, have a substantial visual tectum but a poorly developed olfactory bulb, whereas in dogs and many other animals the olfactory area is large, reflecting a great sense of smell that humans lack.

During evolution from fish to mammal or bird, there have been three main changes in the vertebrate brain. First, the brain simply becomes larger in proportion to body size. Birds and mammals, for example, have brains that are on average up to ten times larger than the brain of a reptile of the same body size. Second, the complexity of nervous connections inside the brain has increased, meaning the bird brain can process much

BRAIN EVOLUTION

The brains of fish, amphibians, reptiles, birds, and mammals look very different, but they are all built from the same four basic components. The "thinking" part of the brain, the cerebrum, is large in mammals and birds but small in the others. This is what we regard as the seat of intelligence. The olfactory bulb is for the sense of smell, the optic tectum for vision, and the cerebellum is for motor control, such as control of movement.

- ■ Cerebrum
- ▨ Cerebellum
- ▨ Optic tectum
- □ Olfactory bulb

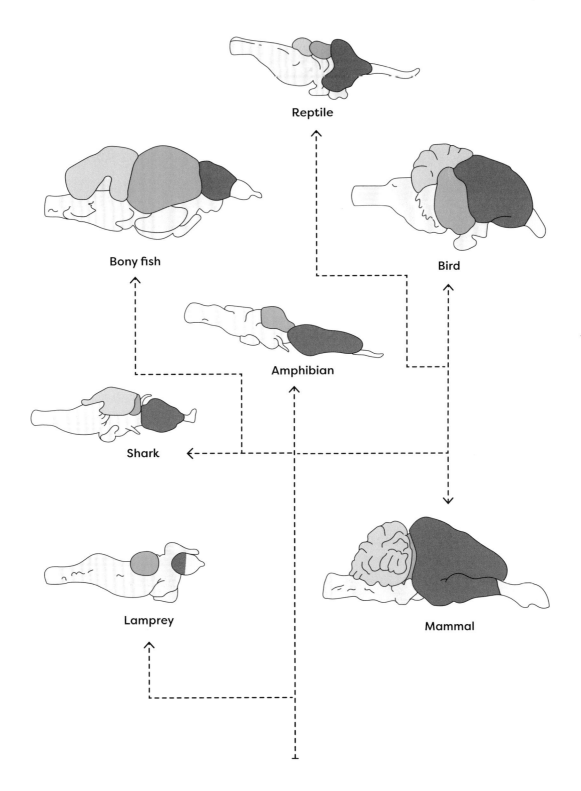

SENSES AND INTELLIGENCE

GOOD EYESIGHT
The early bird *Archaeopteryx* has good vision, as shown by the large eye socket in its skull and a large visual area in the brain. Good eyesight is essential for birds so they can control their flight accurately, especially when landing in a tree, and for chasing fast-moving prey like flying insects.

more information than the fish brain. Third, there are more types of cells in the brains of birds and mammals than in fish brains, meaning they perform a wider variety of functions.

MEASURING INTELLIGENCE

The key to measuring intelligence is the ratio between brain size and body size. To make this clear, whales have bigger brains than humans but this doesn't mean they are more intelligent, just that they are huge. In fact, whales and dolphins are highly intelligent, but we need a more reliable way than just brain size to identify intelligence levels.

The encephalization quotient (EQ) is a commonly used measure of intelligence. This is based on the ratio of brain volume to body volume, and compares the actual brain size to what is expected for an animal of that particular size. The EQ does not capture every detail of intelligence as we see it in humans. For example, human brain size can vary considerably among individuals, but people with big heads (and big brains) are not always smarter than those with smaller heads and smaller brains. However, for animals in general, the EQ gives a guide to intellectual ability.

When biologists compared brain and body sizes for a wide variety of animals, they found there was strong agreement in these measurements for particular groups, and big differences between groups. So it was no surprise to find that insects, fish, and reptiles had relatively small brains and low EQ values, whereas birds and mammals had much higher EQ values. In fact, the jump in relative brain size is by a factor of nearly ten times.

Looking at the graph for mammals, we can see the average line of brain size against body size, and which animals are above and below the line. Above the line are animals we might think of as smart for their size, such as crows, squirrels, cats, monkeys, apes, dolphins, and humans. Less intelligent for their size, below the line, are shrews, rats, pigs, and large whales. On the line are cats and horses.

We can't use the exact figures of brain size versus body size as a hard measure of intelligence, but we can look for whether the EQ is above or below the line of best fit (the straight line on the graph). So, the EQ for humans is 7.44, dolphins 5.31, dogs 1.17—all above the line, but the horse comes in at 0.98 and the mouse at 0.5, some way below the line.

In a classic study, Harry Jerison, an expert on brains and intelligence, showed that most dinosaurs had brains definitely below the line and in the zone of modern reptiles, except for the first bird, *Archaeopteryx*, whose brain size was on the borderline of modern birds.

These measurements are crude because they only look at total brain volume and do not distinguish the parts of the brain. So, an animal with amazing eyesight might have a large visual area and so a large brain volume, but its forebrain, where intelligence resides, might remain small.

ENCEPHALIZATION QUOTIENT

Large animals have large brains, but the relative brain size tells us something about the "intelligence" of an animal. If the brain size lies above the line, as in mice, chimpanzees, humans, and dolphins, the animal is deemed to be more intelligent than expected. On the other hand, animals below the line, such as bats, hedgehogs, pigs, and blue whales, are actually less intelligent than expected.

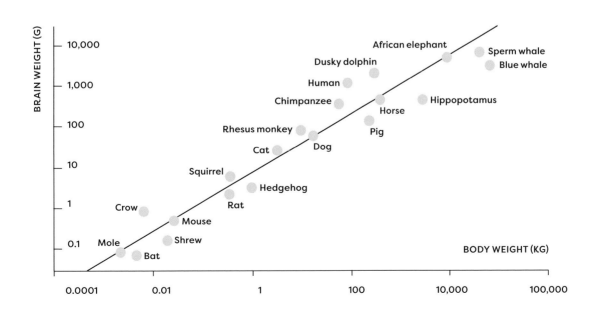

SENSES AND INTELLIGENCE

EVOLVING INTELLIGENCE

Surprisingly, some dinosaurs were probably quite smart, even though most did not have great thinking power. Most of the advances in intelligence happened in the line to birds.

The smartest dinosaurs were the theropods close to the origin of birds. We commonly say that someone is "bird brained" if we mean they aren't very bright, but in fact modern birds have good levels of intelligence, close to mammals. So, in the evolution of dinosaurs, the theropods with feathers (see page 40), many of which were experimenting with different kinds of flight (see page 98) already had a good level of intelligence. We can use the encephalization quotient (EQ) as a way to compare dinosaurs with modern animals (see page 112).

In a study of the dinosaurs on the evolutionary line to birds, Amy Balanoff and colleagues were able to show step-by-step how the transition happened. For a long time, only *Archaeopteryx*, the first bird, had been studied in any detail, and its relative brain size was midway between dinosaurs and birds. So, was there a jump in brain size to mark the origin of birds? In fact, this is not what Balanoff found. She looked at CT data for a number of theropod dinosaurs and found that brain expansion had begun earlier than the origin of birds. She explained this by the fact many of these pre-bird theropod dinosaurs were probably already flying, and so needed the brain complexity to manage the new mode of locomotion as well as cope with complex and diverse habitats and 3D vision. Birdlike brains appeared before birds evolved.

How do the different dinosaur groups compare? Theropods and birds have the highest EQ values, perhaps reflecting the fact that hunting animals need to be smart to catch their prey, so all theropods, even monsters such as *T. rex,* had to have relatively higher intelligence and better senses than the herbivores. Bottom of the heap are the sauropods, which were huge, of course, and so their brains were physically big, but allowing for their large sizes they had low EQ values. Next are the armored dinosaurs, the stegosaurs and ankylosaurs, then the horn-faced ceratopsians such as *Triceratops*, and the ornithopods.

There isn't enough information yet to say whether Cretaceous dinosaurs were smarter than Jurassic dinosaurs, but in one group, the tyrannosauroids, the EQ value did increase and they evolved an improved sense of hearing.

BRAINS OF DINOSAURS AND BIRDS

In the evolutionary tree below, we see how dinosaurs such as *T. rex* and Troodontidae evolved into birds (Avialae, Aves). The brain changed enormously, from dinosaurs such as *Citipati osmolskae* (A) and an unnamed troodontid (B), through the first bird, *Archaeopteryx* (C), to modern ostrich (D) and woodpecker (E). The impressions of the brain (endocasts) of the fossils fit the spaces marked in the sectioned skull of a modern red-tailed tropicbird (F).

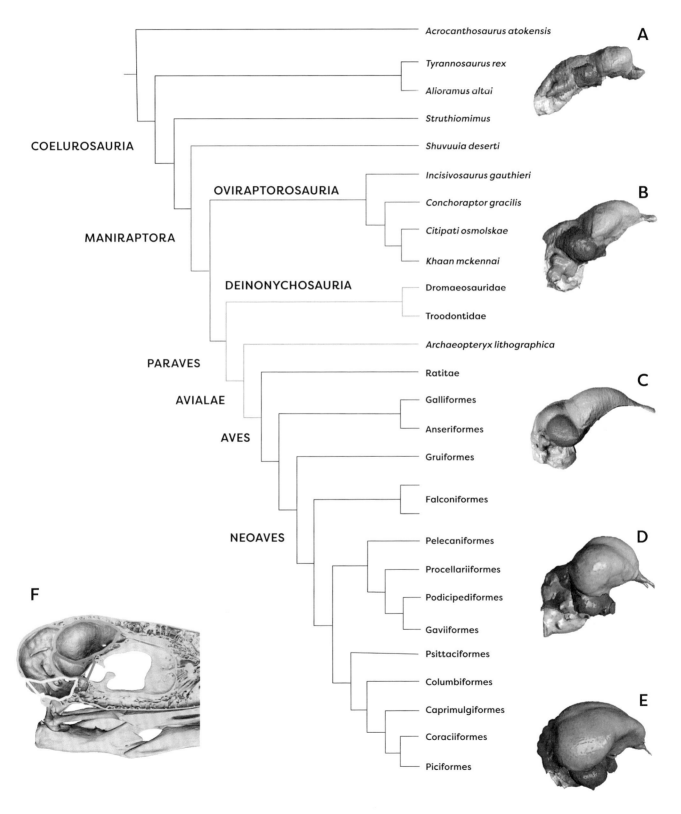

SENSES AND INTELLIGENCE

THE SMARTEST DINOSAUR

How far could dinosaurian intelligence go? In terms of EQ measures, the brightest of all was *Stenonychosaurus*, a long-limbed, 11 ft (3.5 m) long, slender predator with a long neck and a long head. It is best known from the Late Cretaceous of North America. In life, *Stenonychosaurus* was presumably covered with feathers, possibly with a broad feathery tail, judging from the feather coverings of its older relatives from China. But this dinosaur was not a flyer; it was a running predator.

Stenonychosaurus had about the largest brain of any known dinosaur. Compared to a body mass of 88 lb (40 kg) and a brain weight estimated at 1.3–1.6 oz (37–45 g), its EQ is 0.24–0.34, according to Dale Russell (see opposite), which is in the range of guinea fowl and bustards among birds, and armadillos or opossums among mammals.

Why so brainy? *Stenonychosaurus* was a ground hunter, but it had evolved from ancestors that, in the Late Jurassic and Early Cretaceous, were among the animals that experimented with flight. In doing so, they not only evolved complex feathers and color patterns, but also became used to living a more complex life, in the trees. They also evolved binocular vision, so they were capable of seeing in three dimensions (see page 122), essential for a tree-climber or flyer so they could calculate distances and land safely. As a ground-dweller, *Stenonychosaurus* still had huge eyeballs, and presumably excellent sight.

So, the visual and motor parts of the brain were expanded, along with the thinking parts of the forebrain.

So far, nobody has made a 3D model of the brain of *Stenonychosaurus*, but Larry Witmer's model of the brain of *Troodon*, a close relative, shows a regular hindbrain and midbrain, but the forebrain is expanded in comparison to most other dinosaurs, showing the predicted extrasensory and thinking abilities expected.

In 1982, Dale Russell (1937–2019), who was then curator of dinosaurs at the Canadian Museum of Nature in Ottawa, did a thought experiment, and wondered what would have happened if the dinosaurs had not died out at the end of the Cretaceous. In particular, he wondered whether any of the dinosaurs might have increased their intelligence.

Russell imagined a modern troodontid, with the features of *Stenonychosaurus* but having evolved for another 66 million years, and with an expanded brain size of 1100 cm^3, comparable to a human brain. Russell's Dinosauroid walked upright, was about the size of a slender teenaged human, had lost its tail, and had humanlike intelligence. Would this really have happened? Probably not—there is no reason to suppose a smart killing machine of a dinosaur would ever have become like a human.

LEFT
The super-smart troodontid *Stenonychosaurus* from the Late Cretaceous of Canada is a fast-running hunter that has long limbs and short, feathered arms. It has large eyes, excellent eyesight, and a large brain to help control its vision and balance.

RIGHT
An imagined Dinosauroid, if the troodontids had survived to the present-day, with an older view of *Stenonychosaurus* behind, lacking feathers. Dale Russell speculated how these dinosaurs might have become somehow humanlike if they had not died out.

SENSES AND INTELLIGENCE

THE MOST STUPID DINOSAUR

Poor old *Stegosaurus*! This dinosaur has been identified as the most stupid dinosaur ever since the first studies of dinosaur brainpower. It's been said its brain was the size of a walnut, or that it had a kitten-sized brain in a 7-ton body.

Othniel C. Marsh (1831–99) is to blame. He was one of the great dinosaur hunters of the American Bone Rush, when so many well-known dinosaurs were discovered in the American West, and indeed he named *Stegosaurus* in 1877, and it has been a favorite ever since.

In 1896, Marsh had the idea of comparing brain and body sizes of seven dinosaurs, and he estimated brain sizes from their braincase casts (see page 112), and their body masses from overall size. In the case of *Stegosaurus*, he compared it with the brain of a modern alligator. He noticed first that the *Stegosaurus* brain was only about ten times the size of the alligator brain, but that their body masses differed by one thousand times. This meant then that the *Stegosaurus* brain was one-hundredth of the size it ought to have been if it had the proportions seen in an alligator. Indeed, he found that all seven dinosaurs he considered had a relatively smaller brain size than in modern crocodiles. Marsh concluded, "*Stegosaurus* had thus one of the smallest brains of any known land vertebrate."

Not only did Marsh show that *Stegosaurus* had the tiniest brain imaginable, but he also suggested that it might have had more thinking ability in its backside. He followed the line of the spinal cord, the extension of the

nervous system running from the back of the brain inside the vertebral column. The spinal cord gives out nerves to left and right all the way along the backbone, essentially providing the means to operate the whole body using instructions from the brain. Marsh found that the spinal cord in *Stegosaurus* expanded substantially in the hip region, where large nerves passed out to the great muscles of the legs and tail. In life, this was part of the nervous control center to enable the animal to walk and move its tail. The spinal cord swelling in the hips was twenty times as large as the brain, and this suggested the poor beast had more thinking power down there than inside its head.

This, of course, is incorrect, as the only functions of the spinal cord in the hips were motor, meaning they controlled movements and reactions. All the actual thinking happened inside the brain in the head of *Stegosaurus*, but probably there wasn't much going on in there beyond instant reactions to opportunities and threats: "There's some food, go forward"; "There's a bad guy, run . . ."

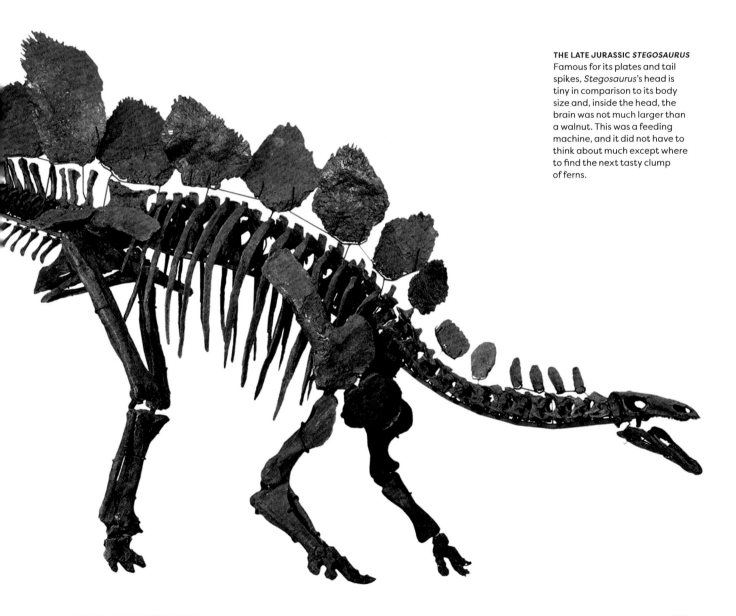

THE LATE JURASSIC *STEGOSAURUS* Famous for its plates and tail spikes, *Stegosaurus*'s head is tiny in comparison to its body size and, inside the head, the brain was not much larger than a walnut. This was a feeding machine, and it did not have to think about much except where to find the next tasty clump of ferns.

SENSE OF SMELL

Humans barely use their sense of smell but for most animals it is key, and dinosaurs evidently had an excellent sense of smell, as shown by their noses and their brains.

We sense smells through our noses and mouths. As humans, we sometimes separate smells that we detect through our noses, and tastes that we detect with receptors on our tongues. But in fact, these sensory systems are linked, and we also smell partly through our mouths and partly through our noses. The sensing is done through nerves that connect to the olfactory bulb of the brain (see page 110).

Most animals have a much better sense of smell than humans. For example, dogs have a sense of smell that is as much as 10 to 100,000 times as good as the human sense of smell. Dogs have 300 million scent receptors in their noses, compared to 6 million in humans. The olfactory lobe in the brain of dogs is forty times larger than ours. It's no wonder that dogs can be trained to use their amazing sense of smell to identify individual humans, or even to detect human diseases, such as cancer, diabetes, tuberculosis, and malaria, from smell alone.

Crocodilians also have an excellent sense of smell, maybe not as good as dogs but much better than ours, and they can smell blood or prey animals in the air and under water. It's generally assumed that birds do not have a

good sense of smell, and that their sense of vision is much better. However, some recent research has shown that some species of birds at least use their sense of smell to navigate, forage, or even to distinguish individuals.

As for dinosaurs, they all had large nasal cavities inside their snouts, and these had multiple functions. Indeed, some, such as in ankylosaurs, twisted and turned in a crazy manner, so when they breathed in and out, the air had to travel a long distance. As it passed through, the dinosaur sensed any odors. The twisting and turning is for heat conservation: as the dinosaur breathed in, the cool air passed over skin lining the passages and picked up warmth from the blood flow, and as it breathed out, warmth was taken back into the body. Humans do this too as an energy-saving device.

We can also assess the sense of smell by looking at the relative size of the olfactory bulb in the brain. In a graph of relative olfactory bulb size against body size, the dinosaurs with the best sense of smell turn out to be some large sauropods and theropods, whereas a poorer sense of smell is seen mainly in smaller and earlier theropods. The early dinosaur *Buriolestes* had a poor sense of smell but great eyesight.

SNIFFING OUT DINNER
These two *Giganotosaurus* predators sniff the air, sensing a tasty meal nearby. The carcass of a dead juvenile *Limaysaurus* gives off foul odors that even we could smell, but the dinosaurs, like modern flesh-eaters, could probably have picked them up from a mile away. The dinosaurs, like modern crocodilians and birds, sampled the air inside their noses and mouths, and they could identify many different smells, some of them especially interesting as they helped them identify their next meal.

SIGHT

Many dinosaurs had an excellent sense of vision, as shown by their large eyes and the large visual area of the brain. Many could see in color, but not all could see in 3D.

Humans are so used to seeing in color and three dimensions that we sometimes forget that this is not true for all animals. Most obvious is that many animals have a long snout, and this means the eyes look mainly sideways, with the forward field of vision divided by their long nose. The secret of 3D, or binocular or stereoscopic, vision is that the field of view of both eyes overlaps, and that's what makes our vision 3D. You can experiment by holding up a finger a foot or so in front of your face, and then looking at it with first one eye, then the other.

So, a horse or dog mainly sees the world on each side with a single eye, and the area of overlap in front of the snout where it has binocular vision is quite narrow, with a blind area just in front of its nose. Probably many dinosaurs saw the world in the same way, and many of them may not even have had that narrow band of binocular vision at the front. Most of the predators at least had the vision range of a horse, though, with a narrow 3D zone at the front, necessary if they were hunting moving prey. Large herbivores that spent most of their time eating plants may not have had even that ability.

FIELD OF VISION

Many theropod dinosaurs had 3D vision. Each eye looked mainly sideways but partly forward, and saw a different view of the world. Right in front of the snout though, the vision fields overlapped, and that's where they saw in three dimensions, essential for a hunter coming in for the kill.

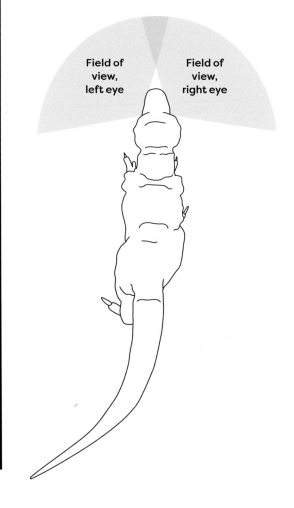

SEEING IN 3D
T. rex almost certainly had good 3D vision. Its eyes are on the sides of a long snout but turned slightly forward, and the snout is actually quite narrow, so *rex* can look straight at you. Three-dimensional vision allowed this hunter to judge distance accurately and to know just how far to jump forward to sink its jaws into flesh.

Some small theropods had very large eyes. Eyeball size doesn't necessarily mean their eyesight was fantastic, but it has certain advantages. For example, among birds, ostriches have the largest eyes, with the eyeball 2 inches (5 cm) across. This is mainly because the eyeball is proportional to body size, and ostriches are large and so have large eyeballs. Among dinosaurs, *Stenonychosaurus*, with its super-large brain (see page 116) also had one of the largest eyeballs, at 1.7 inches (4.4 cm) across.

The large eyeball meant that *Stenonychosaurus*, like most other small theropods, had excellent binocular vision, because the eyes were facing mainly forward.

The large eye could also help it see better in the dark or at times of low light. In modern animals for example, owls have large eyes so they can pick up even tiny amounts of light as they hunt at night. Was *Stenonychosaurus* nocturnal? Possibly.

Did dinosaurs see in color? Modern birds and reptiles can see in color, so dinosaurs probably could as well. Evidence comes from the bright colors and patterns of their feathers (see page 166) which were probably only that way so other dinosaurs could see them. Unexpected new evidence from molecular biology seems to confirm that dinosaurs had color vision.

SENSES AND INTELLIGENCE

HEARING

Dinosaurs had good hearing, and paleontologists can detect the kind of hearing from the brain and the semicircular canals in the inner ear. *T. rex* could hear low sounds very well.

Among modern reptiles, crocodiles and most lizards hear well, but snakes and turtles have a poorer sense of hearing, some of them relying on picking up vibrations through the ground rather than sounds through the air. In fact, the inner ear structures allow ears to "hear" in both ways.

Many people think reptiles and dinosaurs had no ears. However, what we call an ear in humans and mammals is the pinna, or ear flap, the external structure that gathers sound. The hearing part of the ear is buried in the skull. This consists of an eardrum and some

NIGHT-HUNTER
The early hunter *Shuvuuia* prowls for insect prey at night. This small predator had an amazing structure in its inner ear just as in modern owls, and this shows it could hunt at night, using its super-sensitive hearing to hear even the slightest rustle as a cockroach hurried along under the leaves.

ear ossicles, tiny bones that transmit sound from the eardrum to the brain.

In all animals, sounds reach the eardrum, which is a layer of skin stretched across the hearing canal. Sounds through the air or ground make the eardrum vibrate, and these vibrations are picked up by the ossicles. Reptiles and birds, including dinosaurs, have a single ossicle, called the stapes. Mammals also have the stapes, but they have two further ossicles, the malleus and incus, or hammer and anvil. This gives mammals more refined hearing, and it seems likely that reptiles do not rely so much on hearing as we do. The ossicles pass the sounds to the cochlea, attached to the semicircular canals (see page 127), then through a nerve to the brain.

In a 2021 study, Jonah Choiniere of the University of the Witwatersrand, South Africa, and colleagues showed that many theropods had the same hearing abilities as owls. They looked at the structure of the inner ear, and especially the semicircular canals. The lower part of this structure, the lagena, is involved in hearing. In modern reptiles and birds, the lagena has a role in balance (see page 126) but it also functions in hearing: the longer the lagena, the better the hearing.

Modern owls and other night birds have a long lagena, and it turns out that this is true for many theropods. Most extreme is *Shuvuuia*, a small Late Cretaceous theropod from Mongolia, which has an owl-like lagena. Coupled with its large eyes, it is likely *Shuvuuia* was nocturnal, seeing in the gloom and hearing the slightest of sounds. *Shuvuuia* was an alvarezsaurid, a group of theropods with tiny, punchy arms that might have fed on ants and termites (see page 161), and being able to hunt at night perhaps allowed it to prey on insects that were nocturnal, being active when temperatures were not so hot. Owls can hear the rustle of a moth half a mile away, and perhaps some small theropods had the same amazing ability.

TESTING FOR NOCTURNAL BEHAVIOR

These 3D models are laser-printed from scans of the inner ear anatomy of a modern barn owl (left) and the tiny hunting dinosaur *Shuvuuia* (right). The overall shape is the same, especially the long structure at the base, called the lagena. The lagena helps animals to keep balance, and it is longest in owls so they can fly and walk at night in total darkness. The same must have been true for this amazing small dinosaur.

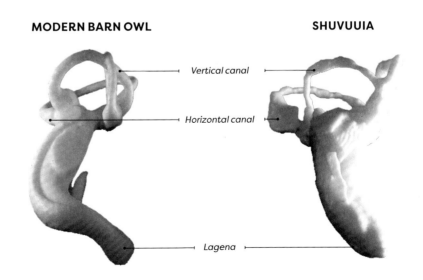

MODERN BARN OWL　　　　**SHUVUUIA**

Vertical canal

Horizontal canal

Lagena

SENSES AND INTELLIGENCE

SPATIAL ORIENTATION

The semicircular canals inside the ear are key to keeping balance, and these were well developed in dinosaurs, and show us how they stood, moved, and balanced on one leg.

Our sense of balance is controlled by the semicircular canals, remarkable structures that are deep inside the ear canal and form part of the structure called the labyrinth, which includes the cochlea. In humans, there is a deep ear canal that ends in the eardrum. The three ear ossicles (see page 125) transmit sound to the cochlea, where the large cochlear nerve transmits the vibrations to the brain for interpretation.

The semicircular canals form a unit with the cochlea. There are three semicircular canals, two of them standing vertically, and one horizontally. The whole system is filled with fluid and there are sensory hairs to detect movement of the fluid. The horizontal semicircular canal detects movements of the head when it turns from side to side, and the two vertical canals detect up-and-down movements. As the head moves, the fluid inside the system moves in response, and the detector hairs transmit the information to the brain.

The semicircular canals function to tell us automatically how our head and body are oriented. Even with our eyes shut, we know if we are upside down, lying, or twirling in circles.

The information is important in helping us stay balanced—when you ride a bike or skateboard, your body is turning and twisting and you would fall over if your muscles didn't react to keep balance, through instant messages via the semicircular canals, through the brain processing, and to the muscles.

We assume that dinosaurs did not ride bikes or skateboard, but the bipedal ones at least were in constant danger of falling over. As they, and you, walk, you generally have just one foot on the ground as the other foot swings into the next step. The whole body has to swing from left to right as you walk or run, and this keeps your balance. This fine adjustment operates through the balance system. Some dinosaurs, such as *Velociraptor* and *Deinonychus*, had large claws on their toes and so may have balanced on one leg while they slashed with the other. This required the balance system of a dancer.

The semicircular canals orient the head correctly when an animal is walking. In a biped the head is on top of the backbone and looks forward, but in a quadruped the neck is more horizontal and the head tilts down to the ground. The Early Cretaceous ceratopsian *Psittacosaurus* showed a remarkable posture shift with growth (see page 189), from quadrupedal baby to bipedal adult. In a 2019 study, Claire Bullar from the University of Bristol, UK, showed that the semicircular canals of this dinosaur shifted from downward to more horizontal as it grew older, matching the posture shift.

PSITTACOSAURUS GROWTH AND POSTURE

The early ceratopsian *Psittacosaurus* started life as a quadruped, and as it grew it became more and more bipedal (see silhouettes of posture, from A to C). At the same time, the posture of its head switched from tipped down and looking forward as a baby (A), to more horizontal as a teenager and adult (B, C), and the orientation of the horizontal canal of the semicircular canals (shaded pink in these 3D skull models from CT scans) proves this: the angle reduces from 38° through 25° to 15°.

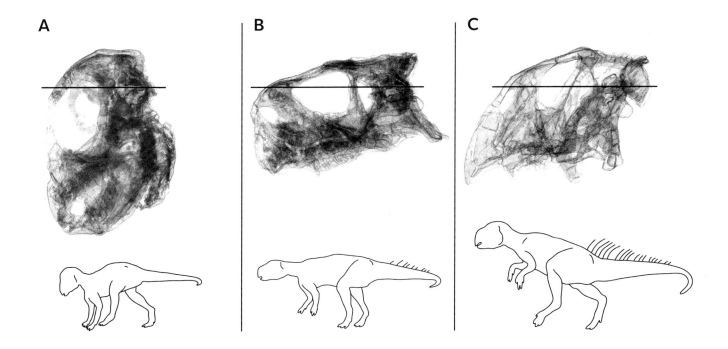

SENSES AND INTELLIGENCE

TYRANNOSAUR SENSES

Everyone wants to understand *T. rex*. Here was one of the largest predators of all time, and it is fascinating to determine whether it was a hunter or scavenger (see page 158), how fast it could run (see page 89), and how it found enough food to power its huge body. The senses—smell, vision, and hearing—can help us understand how tyrannosaurs survived.

All the evidence suggests that most theropods, including the tyrannosaurs, had a good sense of smell (see page 120). In reconstructions of the olfactory bulbs of the brain (see page 121) these were relatively large, indicating a better sense of smell than many other dinosaurs. Tyrannosaurs presumably used this sense of smell to sniff out prey animals, whether dead carcasses or living beasts.

We have seen that large theropods such as *T. rex* had good vision, especially some binocular vision straight in front of the head (see page 122). In life, *T. rex* probably tilted its head down by 5 or 10 degrees, which would have allowed the eyes to see forward better and broadened the field of 3D vision, important when focusing in on a prey animal. In fact, with its head down, the field of binocular vision in front of *T. rex* was as wide as 55 degrees, as good as a modern hawk.

The inner ear structures suggest that *T. rex* heard low-frequency sounds especially well, such as the grunting and rumbling of herbivorous dinosaurs.

The semicircular canals in *T. rex* are unusually long, which indicates a good sense of balance. They could also suggest that *T. rex* moved fast, but it's more likely that they enabled the predator to keep its head and eyes fixed on its prey, even when it was running or turning. Gaze-fixing is seen in hunting birds, which lock onto their prey from some distance, and keep a laser-like visual link through all their twists and turns as they prepare to swoop in and snatch their victim. For *T. rex* this would have been crucial in hunting so that the prey could not jump sideways or hide behind a tree.

So, it probably wasn't like the *Jurassic Park* movies, in which it's said you can outwit *T. rex* on the prowl by standing still. You could stand there as still as you like, but the *T. rex* would lock its eyes onto you, rush over, and snap you up, thanks to its great balance and eyesight. As Larry Witmer at Ohio University concluded, "Tyrannosaur sensory biology is consistent with their predatory . . . heritage, with emphasis on relatively quick, coordinated eye and head movements, and probably sensitive low-frequency hearing; tyrannosaurs . . . enhanced their olfactory apparatus."

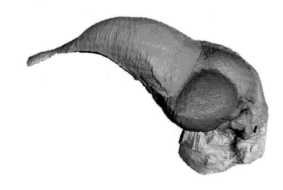

THE BRAIN OF THE FIRST BIRD, *ARCHAEOPTERYX*
From front (left) to back (right) the main brain areas are colored: olfactory area (orange), cerebrum (green), optic lobes (red), cerebellum (blue), and brain stem, connecting to the spinal cord (yellow).

T. REX CRANIAL ENDOCAST

In life, the entire brain of *T. rex* was smaller than the human brain. Relative to its huge size, weighing over 5 tons, this is a puny brain and indicates very limited intelligence. The fossils, though, show amazing detail of the brain, the cranial nerves (yellow) and some blood vessels (blue). The brain confirms excellent senses of smell (large olfactory bulb) and vision (large optic lobe), and good control of limbs and jaws (large cerebellum).

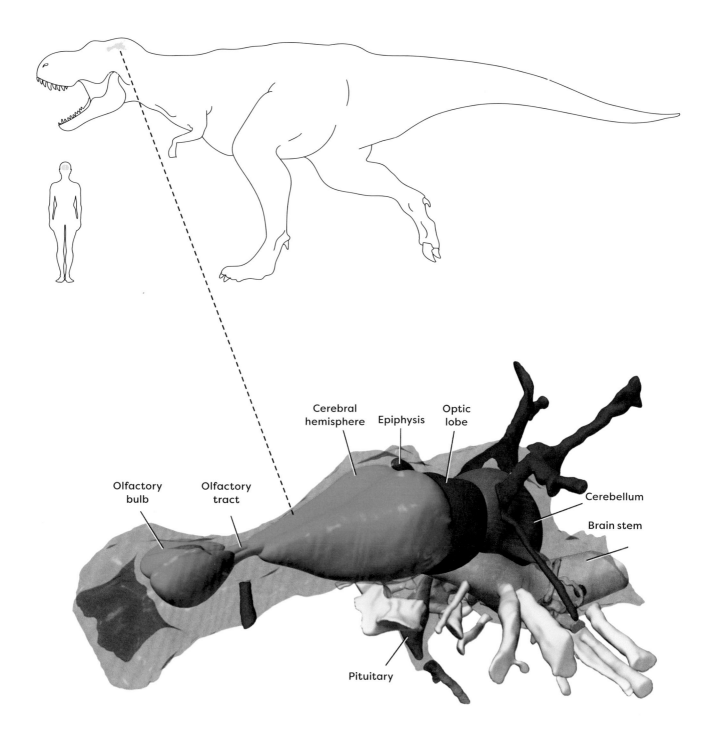

SENSES AND INTELLIGENCE

BIRD BRAINS AND SENSES

Bird brains are squeezed into the back of their skulls instead of extending along the whole skull roof as in most dinosaurs and other reptiles. However, the bird brain has an enlarged forebrain, the thinking part, as well as a large visual area. These features are seen in the brain of *Archaeopteryx*, the first bird, which already had the scrunched-up brain position and large areas for the senses of vision, hearing, and balance. The expanded forebrain matches our assumption that *Archaeopteryx* could already fly (see page 102) and needed considerable thinking ability to cope with its complex world.

As we saw earlier, birdlike brains evolved before birds did. This indicates that many of the maniraptoran theropods, dinosaurs such as oviraptorosaurs, dromaeosaurids, troodontids, and anchiornithids, were experimenting with complex lifestyles in the trees, and many were indeed flying and gliding (see page 101).

After that, bird brains continued to develop, with expansions of the forebrain and the optic lobes, perhaps reflecting improvements in maneuverability in flight. But the brain of *Ichthyornis*, one of a group of birds that went extinct at the end of the Cretaceous, is not as birdlike as the brains of birds that survived the extinction. In these modern forms, the cerebral hemispheres (forebrain) are much expanded, indicating they were much smarter.

Do dinosaur brains show other changes in flyers? The labyrinth certainly does. In birds, the semicircular canals in the labyrinth are more firmly at right angles to each other, and each canal is more circular. The canals are different shapes, and the canal lying to the front in particular has a large, open looplike shape.

RELATIVES FOR DINNER
Brainy *Velociraptor* starts to eat a close relative, *Mahakala*. These small predatory *Mahakala* used their excellent vision, smell, and hearing to hunt or to escape from hunters. This was a kind of arms race, where predator and prey species evolved to improve their senses.

In terms of overall canal shapes, there is a kind of evolution of shape from the earliest ancestors of dinosaurs: through a line to crocodilians in one direction; then dinosaurs, pterosaurs, and relatives; then birds seem to show a separate set of features. Unexpectedly, many dinosaurs show birdlike features in their labyrinth system, which is overall very large in proportion to body size. This large size is seen in modern birds, even those that no longer fly, and it occurs also in most dinosaurs and in some pterosaurs.

It's hard to link any of these changes directly to flight because they occur both in flying birds and in nonflying dinosaurs. Perhaps the overall large labyrinth and the rearranged semicircular canals relate to functions of both ground-dwelling dinosaurs and flying birds, such as good balance, agile running, and gaze-fixing in predators.

SEMICIRCULAR CANALS AND FUNCTION

The semicircular canals from the ears of dinosaurs can be used to indicate function when they are compared with modern animals. Here, the complex shapes are coded along two axes, mapping the shape. Among modern reptiles, the semicircular canals of turtles, lizards, crocodilians, and birds fall in different zones of the morphospace, indicating different shapes, generally connected with different functions.

Note: The x and y axes represent a summary of shape variation between all the semicircular canals: the x-axis corresponding to most shape variation (57.6% of the total); the y-axis to the second-most amount of shape variation (11.1% of the total).

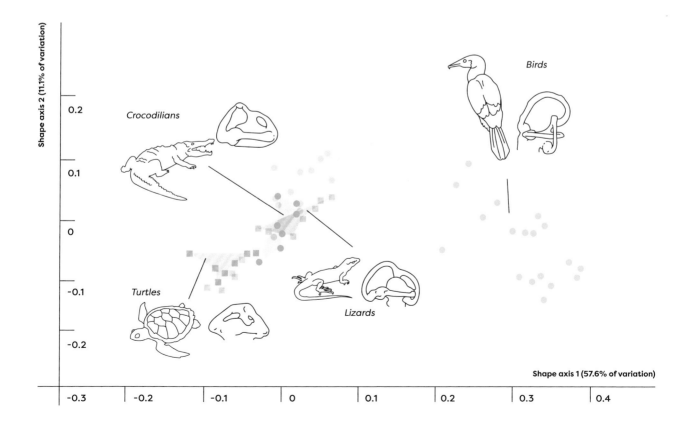

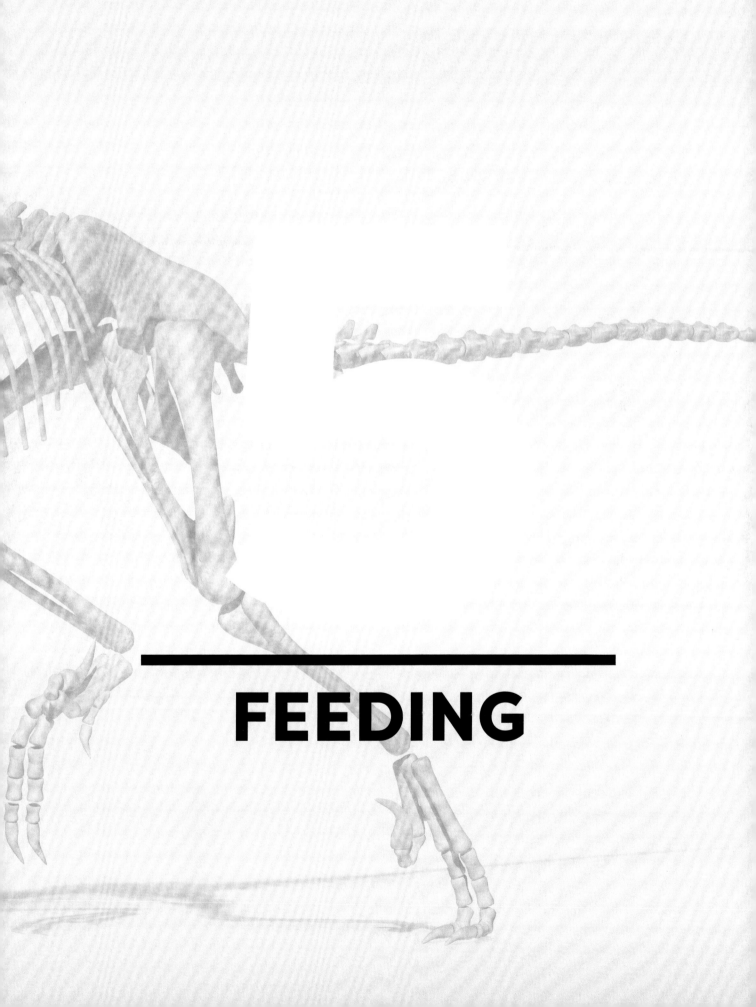

FEEDING

DINOSAUR DIETS

Dinosaurs fed on resources from the landscape, including plants, insects, other animals, and one another. Here we identify dinosaurian herbivores, carnivores, and omnivores.

It's important to determine what each dinosaur would typically eat in order to understand their ecology, including food webs and food pyramids (see page 69). This requires study of body size and relative abundances of different dinosaur groups at different times. It's well known that the theropods were mainly carnivores, and all the others, the sauropods and ornithischians, were herbivores. But it's not quite as simple as that, as many theropods, such as therizinosaurs and oviraptorosaurs, became herbivorous; others such as alvarezsaurids might have been specialist insect eaters, and some early dinosaurs were omnivores, apparently happily eating meat and plants.

How can we tell? Evidence about diet comes from the teeth and jaws of dinosaurs. But it also comes from the study of stomach contents and particularly dinosaur poo. In addition, paleontologists now use computational methods in biomechanics to explore diet and feeding, and the chemistry of the bone can even provide clues about diet.

In modern animal communities, the largest herbivore is usually larger, or much

SIZE RANGES OF PREDATORS (of the Hell Creek Formation in the Late Cretaceous of North America). Largest is *Tyrannosaurus*, then there is a gap in size to the smaller predators such as *Acheroraptor* and *Leptorhynchos*. But between these there was a size gap, with no medium-sized flesh-eaters. Probably the gap was filled by juvenile tyrannosaurs.

larger, than the largest carnivore. For example, in Africa, elephants can weigh 5–6 tons, and the largest predator, the lion, weighs 400 lb (190 kg). The same is true for most fossil communities of mammals, too. But for dinosaurs, the size ratio was variable, starting distinct and ending up much closer. This was because the sauropods were the key herbivores in the Jurassic, and they weighed 10–50 tons; their predators, such as *Allosaurus*, weighed a mere 2 tons. But sauropods disappeared in many locations in the world, and in the Late Cretaceous there was less difference in size between, say, a hadrosaur and a tyrannosaurid predator, both weighing 2–4 tons.

These size differences mean that in the Late Cretaceous, carnivores such as *T. rex* were presumably active hunters because there is a good match between the average sizes of herbivores in a typical community in North America with carnivore size ranges. Not only do the size ranges match between herbivores and carnivores, but so too do the relative abundances. That confirms that, as today, predators hunted prey of about their own size.

THE CARNIVORE GAP

It seems reasonable to assume that any animals in an ecological community will show a complete range of body sizes from small to large. Among carnivorous mammals, for example, there's a complete range from tiny weasels, through hunting dogs and hyenas, all the way up to tigers and lions. Roughly speaking, the size of the predator determines the size of the prey they can tackle.

In 2021, graduate student Kat Schroeder at the University of New Mexico noted something strange about theropod sizes in many fossil communities. There were plenty of small ones—all the little feathered coelophysids, dromaeosaurs, and oviraptorosaurs—weighing less than 130 lb (60 kg), and plenty of large

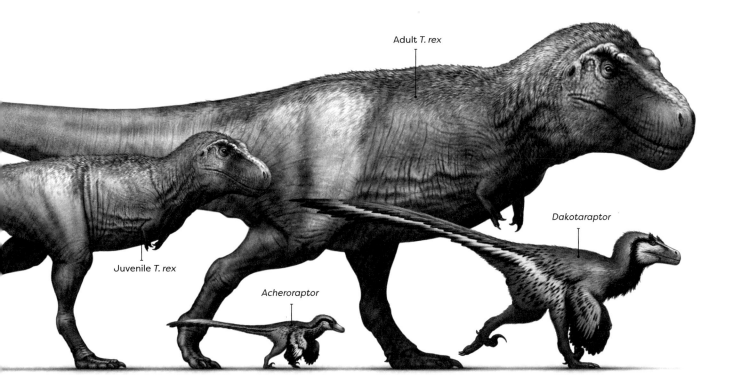

SIZE MATTERS
From the Late Jurassic Morrison Formation in Montana, the adult *Barosaurus* is so large it cannot be attacked, but the predator *Allosaurus* is able to kill and eat a juvenile *Barosaurus*.

ones—the allosaurids, sinraptorids, abelisaurids, carcharodontosaurids, and tyrannosaurids—weighing typically more than a ton.

But what about the missing ones in between these size ranges? The researchers showed that the absence of middle-sized predators was not because middle-sized herbivores were also absent. In fact, comparing size ranges of herbivorous and carnivorous dinosaurs, they saw the difference clearly: there were plenty of herbivores of all sizes, but a definite gap in carnivore size. What was eating the medium-sized prey animals?

Schroeder and her colleagues argued that the missing predators in the 200 lb–1 ton (100–1,000 kg) range were the juveniles of the giants. It took five to fifteen years for a large theropod to reach adult size (see page 188), and that gives the teenagers a few years plugging the medium predator gap.

The researchers noticed a change from Jurassic to Cretaceous, with the gap widening through time. In the Jurassic, the teenaged predators were more like their parents, whereas the Late Cretaceous predators were larger as adults, and so took longer to reach adult size, and the juveniles could plug the wider size gap.

This kind of study suggests a new way of thinking about dinosaur growth. Unlike mammals, where the babies stick around with the juveniles and adults, and there isn't a huge change in size from baby to adult, it was different in dinosaurs. It's all because they laid eggs, which were generally small in comparison to adult size (see page 176), so the hatchling dinosaurs probably kept apart from the adults, living their own lives for safety. In the long journey from small baby to huge adult, they passed through a whole range of feeding modes, each step along the way acting like a different species.

This is one reason why maybe the total number of species in typical dinosaur communities was a lot lower than might be expected: each growth stage within a single species perhaps acted like a distinct species in terms of its place in the food chain.

TEETH AND JAWS

Carnivorous dinosaurs had sharp, curved teeth with serrated edges, whereas herbivores had blunter teeth, sometimes pointed, often not, but suited for chopping, not chewing.

It's usually easy to tell whether a dinosaur was a plant-eater (herbivore) or flesh-eater (carnivore) by the shape of the teeth. In fact, this is true of all vertebrates. We know, for example, that cats and dogs have sharp teeth, whereas cows and horses have blunt teeth. Humans have mixed tooth types, including sharp incisors and canines for grasping food and piercing meat, and blunt molars at the back for chewing.

Most reptiles do not show this kind of differentiation of teeth—incisors, canines, and molars—as seen in mammals. Mostly, they have the same teeth from front to back of the jaws. When we look at a dinosaur tooth it is divided into a crown, the part above the jawline, and a root, the part embedded in the jaw.

In dinosaurs, carnivore teeth were sharp-pointed, usually curved back, and the front and back margins were lined with serrations, a series of small spikes like the blade of a steak knife. The backward curve of the teeth is common in flesh-eaters because it effectively pushes the prey down into the throat; if the prey animal struggles, it impales itself further onto the pointed teeth and moves further back in the mouth.

Herbivorous dinosaurs showed a great range of tooth types. Some herbivores had leaf-shaped teeth, with deep roots. The crown may be roughly triangular in shape and symmetrical, with broad serrations for slicing vegetation, and sometimes a middle ridge to help divide the stems and twigs. Ornithopod teeth are usually slightly higher than ankylosaur and stegosaur teeth, which can have low crowns. Most sauropods had one of two types of teeth. Some such as *Diplodocus* and *Brontosaurus* had long, pencil-shaped teeth that pointed forward at the front of the mouth and were used to strip leaves from branches. Others such as *Camarasaurus* and *Brachiosaurus* had spoon-shaped teeth with long roots, and their teeth extended all along their jaws.

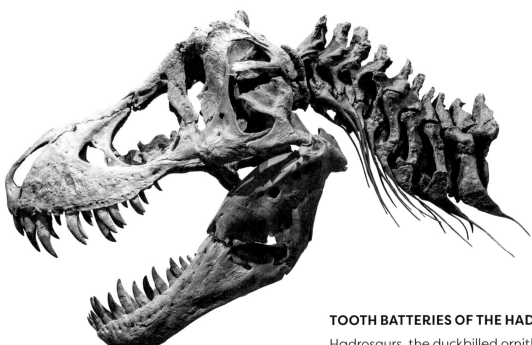

TEETH AND JAWS TELL US ABOUT DIETS
The amazing tooth batteries of hadrosaurs (above) consisted of a series of seven to eight teeth lined up one below another, always growing, and providing a powerful set of teeth for chopping tough plant food. The *T. rex* skull (top) shows the long, curved, sharp teeth typical of a flesh-eater—perfect for killing and for chopping meat.

TOOTH BATTERIES OF THE HADROSAURS

Hadrosaurs, the duckbilled ornithischians of the Late Cretaceous, had the most amazing set of teeth of any herbivorous dinosaur. In fact, unlike most other reptiles, it seems they could chew their food almost in the same way as mammals, meaning they could grind it up between their teeth before swallowing. Chewing is one of the reasons mammals are so successful, because it enables them to extract maximum goodness from the food.

In hadrosaurs, the chewing happened by small movements of the jaw bones from side to side as the jaw opened and closed. These sideways movements enabled the teeth to rip up the plant stems before they were swallowed. Recent studies showed that hadrosaurs had six different materials in their jaws and teeth to make them especially efficient. First, the teeth themselves were composed of four different hard tissues, called enamel and dentine, and two types of cement. Our teeth are also composed of dentine,

MEAT-EATER
The theropod *Majungasaurus* has sharp, pointed teeth for tearing flesh. In this image, only the tips of the teeth show up because it is thought thick lips covered the teeth when the jaws were closed.

covered with a cap of shiny enamel, and fixed into the jaws with cement. In hadrosaurs, the two types of cement were not just for fixing the teeth, but also they had become part of the grinding surface.

There were two other hard tissues in the hadrosaur teeth: filled giant tubules in the core of the tooth, and a second kind of dentine. Altogether, these studies showed that hadrosaur teeth were some of the most complex ever seen. Mechanical experiments showed that the different kinds of dentine, enamel, and cement wore down at different rates, and this left a ridged surface after wear, ready to rip up the next lot of food. These kinds of complex, multi-ridged, and constantly sharpening teeth were known before only in specialist plant-eating mammals such as elephants and cattle.

Hadrosaurs had one big advantage over mammals, though, which was that they showed continuing tooth replacement throughout their lives.

TOOTH REPLACEMENT

As mammals, we think it's normal to replace your teeth only once. Mammals tend to be born with no teeth, then the baby teeth appear, which are in turn replaced by the adult teeth, and that's that. If you don't look after your adult teeth by brushing and good dental hygiene, they rot and fall out. In the wild, mammals do not wear false teeth, so an old deer or elephant can die if its teeth have simply worn down.

This is unusual and found only in mammals. In most fish and reptiles, the teeth replace continuously. As teeth wear down, new ones pop up from below, and indeed some dinosaurs, such as the hadrosaurs, as we have seen, might have a battery of as many as seven or eight replacement teeth lining up in any tooth position, ready to move into place.

For carnivores, this was a real problem, especially if they had to struggle to subdue their prey. Just as with sharks today, it was easy for them to lose teeth, which simply broke off and fell out. In the case of the large theropod *Majungasaurus,* from the Late Cretaceous of Madagascar, close study of growth lines in the jaw bones suggest that it replaced its teeth every two months. We don't know whether this was unusually fast or just normal for large predators.

FORENSICS:
FINITE ELEMENT ANALYSIS OF JAWS

It is possible to test the function and strength of dinosaur jaws, and even answer questions such as: What was the bite force of *T. rex*? Could *T. rex* feed by twisting its head from side to side and ripping back to tear flesh from bone? Paleontologists use some standard methods from engineering, including finite element analysis (FEA) and multibody dynamics (MBD), to answer such tricky questions.

FEA is mainly used in designing large structures such as buildings, bridges, and aircraft, especially in testing how such structures perform when they are made from different materials. MBD is used to design machines with moving parts and to test how they move relative to one another.

Emily Rayfield of the University of Bristol, UK, has applied FEA to many questions about dinosaur jaw function. First, she makes a complete scan of the skull. Scanning can be done either using X-rays, as in computed tomography (see page 108), or surface scans, which are lots of photographs taken from all sides, which are then stitched together like a 3D landscape. This is cleaned up, any distortion of the fossil is removed, and missing parts are restored, sometimes by matching and modeling left to right elements, because skulls are symmetrical (if the right jugal is missing, the analyst can take the scan of the left jugal, flip it, or reverse it, and make a model by matching from left to right.)

When the digital skull model in the computer is complete, Rayfield breaks it up into hundreds or thousands of elements, so it looks like a wire model or mesh of the skull, and applies the material properties—different for teeth, compact bone, open bone, etc. Then she can apply forces that might be experienced during different kinds of feeding.

In comparing three theropod skulls, Rayfield found that *Coelophysis* and *Allosaurus* were most similar, and differed from *Tyrannosaurus* in the locations of zones of maximum stress (shown by hot colors, such as red and yellow). In *T. rex* the snout is most under stress and the bones in this region are fused together to make them stronger. In *Coelophysis* and *Allosaurus* the maximum stress is further back, above the eye sockets. These differences show that each theropod had a different way of biting—faster and weaker in *Allosaurus*, heavier and harder in *Tyrannosaurus*.

MBD calculations confirm that *T. rex* had an amazing bite force of 35,000–57,000 Newtons, or the equivalent of 4–6.5 tons of force.

This is nearly ten times the bite force of the great white shark.

COELOPHYSIS

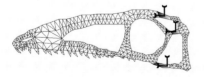

ALLOSAURUS

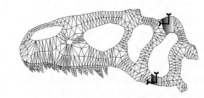

TYRANNOSAURUS

EVIDENCE OF DINOSAUR DIETS

Dinosaur diets can be identified from their teeth, jaws, remains of food in the gut, fossil vomit, and especially fossil poop, but chemical methods can provide new evidence.

How can we identify what a dinosaur was eating? As we have seen, the teeth usually give a broad picture, and FEA study of the jaws can identify how they worked. But what about more direct evidence? We want to see the feeding operation "caught in the act."

Paleontologists use basic ecological information. They look at the plants and other animals found side-by-side with their dinosaurs of interest. For example, if we find specimens of the herbivorous dinosaur *Triceratops*, we look for the evidence of fossil plants, such as fronds of ferns, leaves, stems, and seeds of conifers, and even some early flowering plants such as roses and dogwoods. This shows us what was on the menu, but does not prove that *Triceratops* chose to eat any or all of them.

Direct evidence of diet comes in the form of bite marks, stomach contents, gastroliths, vomit, and poop. Some dinosaur poop shows us parasites that hint at dinosaur diseases. A new source of evidence about diet comes from studies of the chemistry of bones.

A PAINFUL BITE
Here, *T. rex* takes a bite out of the tail of a hadrosaur *Edmontosaurus*, but the hadrosaur eventually escapes, carrying the evidence of the fight—a *T. rex* tooth embedded in the tail vertebra. The wound healed and the hadrosaur lived on for a few more years, but that tooth must have given her a sore tail.

BITE MARKS

Bite marks can provide good evidence of theropod diets. We don't know of any records of identifiable dinosaur tooth marks in leaves or plant remains, but there are plenty of reports of bones with dinosaur tooth marks in them. Sometimes it's possible to work out the shape of the tooth if it went straight in as a puncture mark, or the serration spacing if the tooth raked across the bone. These factors can help identify the predator.

One amazing specimen reported by Robert de Palma of the Palm Beach Museum of Natural History, in Florida, shows a pair of hadrosaur tail vertebrae with the tip of a *T. rex* tooth embedded. The hadrosaur escaped this attack and lived for some time, but the broken tooth caused infection of the vertebrae and a thick bone lesion grew around the site.

There have even been reports of *T. rex* bones showing marks of *T. rex* teeth, and this has been used as evidence they sometimes practiced cannibalism, eating members of their own species.

STOMACH CONTENTS, GASTROLITHS, AND VOMITITES

It's rare to find remains of a dinosaur's last meal inside its stomach. This is because the internal organs, including the guts, are made of soft tissues, which rot away quickly after the animal dies. Also, in most cases the dinosaur skeleton has been disturbed after death by scavengers, detritus-eating insects, and the like, and they love to eat guts and partly digested food. In both herbivores and carnivores, the gut usually fills with gases of decay after death, and it probably explodes. There are cases today of elephant or whale carcasses exploding and blasting surrounding people with lumps of rotten flesh and stinking food.

In one case, though, a remarkably preserved ankylosaur, *Borealopelta*, from the Early Cretaceous of Alberta, Canada, showed its last meal. The large carcass had become trapped in tar sands, which are sands filled with thick oil bubbling beneath the surface, and in sinking into the ground, the oily deposit preserved the flesh.

Borealopelta was an 18-ft (5.5-m) long armored giant, covered externally with an armor of bony plates large and small, and some with sharp scimitar edges. But it was not a warlike animal, feeding, as the stomach contents showed, on fern fronds, with a few leaves of conifers and cycads. Unexpectedly, some of the leaves and twigs had turned into charcoal, meaning they had been burned before being eaten, so this is evidence that there had been some kind of forest fire in its habitat before it prowled across and nuzzled up the plant food it could find. The state of development of the leaves and seeds tell us that *Borealopelta* ate its last meal in spring or early summer, a time of forest fires.

In its stomach also were some gastroliths, literally "stomach stones." Modern birds swallow grit to help digest their food. Birds of course have no teeth, and chickens peck around the farmyard, grabbing seeds and sand grains. The grit passes to a structure in their throat called the gizzard, in which the

STOMACH STONES
The gastroliths inside this rib cage of a Cretaceous waterbird from North China, *Gansus zheni*, are small and rounded, and they helped grind up the plant food inside the digestive system. Perhaps early birds and dinosaurs, like modern birds, had a crop, a special structure that food passed through on its way to the stomach, and where the food was ground up by this swallowed grit.

HISTORIC VOMIT
Is this the first example of dinosaur vomit? The photograph shows a bunch of bones of a lizard-like protorosaur, all crushed and lined up in a way that suggests they had been partly digested, then vomited out perhaps by an early dinosaur.

grains and tough food are crushed by muscular squeezing of the crop and its content of stones. The mush then passes down to the stomach, where it is digested and the goodness extracted. Some crocodiles, seals, and sealions also swallow gastroliths.

Dinosaurs probably did not have a gizzard, but many groups, especially herbivores, seem to have shared with birds the habit of swallowing grit. Well, because dinosaurs were huge, their grit, the gastroliths, were typically several inches across. It's hard to be sure you have found a dinosaur gastrolith as they are not just any pebble found near a skeleton. They have to be rounded and polished by action in the stomach, preserved within the rib cage where the gut would have been.

Did dinosaurs vomit? Presumably they did, and several examples have been noted. These are a specialized kind of vomit called pellets, where a carnivore crunches up the skeleton of a smaller animal, begins to chomp and digest, then vomits up the bones that are indigestible. Today, ornithologists are familiar with owl pellets, where the owl grabs a small mammal, crunches it up, and spits out a ball of fur and bones. An example of a dinosaur pellet from the Late Triassic of Italy contains bones of a lizard-like protorosaur.

More normal puke would not survive as a fossil because it's a liquid mess, but Tony Martin of Emory University, Georgia, calculated the effect if *Brachiosaurus* vomited. With its head 45 ft (14 m) above the ground, a stomach-full would be 110 lb (50 kg), and it would have hit the ground with a force of more than 7 tons. Keep well clear!

FEEDING

SWALLOWING ROCKS
The Early Cretaceous ankylosaur *Borealopelta* collects rock pebbles. It needs these to help it grind up its plant food. As dinosaurs had simple hinge-like jaws, they could not move their jaws from side to side to chew their food, so they swallowed grit (small rocks) to help crush the food and extract the goodness.

COPROLITES

Dinosaur dung is endlessly fascinating. Unlike the other indicators of diet, it is also relatively common. Indeed, some bonebeds, deposits made from masses of bones, usually small bones, may also contain great numbers of individual poos, or coprolites. These may be preserved with rich amounts of phosphate. Historically, some coprolite beds preserved under shallow seawater, and full of coprolites of fish and marine reptiles, were mined as agricultural fertilizer. In Ipswich, UK, there is even a Coprolite Street, where a fertilizer works once stood.

The problem with coprolites is that you cannot always work out exactly who dropped them, even though their great value is that they often contain identifiable fragments of plants or bones. Vandana Prasad of the Birbal Sahni Institute of Paleobotany, India, reported sauropod coprolites from the Late Cretaceous of India. These surprisingly small, pellet-like coprolites suggest the dinosaur dropped masses of ball-like poop like a horse, rather than a big flop like a cow. Inside, unexpectedly, the researchers identified seeds of many kinds of grass. At the time, grass was rare, and not usually seen as food that dinosaurs ate, but the proof was there in the coprolite.

The most amazing dinosaur coprolite was reported in 1998 by Karen Chin, then at the Museum of the Rockies in Bozeman, Montana. This was a 17-inch (43-cm) long monster, "the granddaddy of all coprolites," as she called it. It contained bones of a subadult ornithischian dinosaur, but the species could not be determined. One observation was that the bones of the prey animal had not been dissolved a great deal by the stomach acids, so tyrannosaurs were not capable of completely digesting their prey, as many crocodilians do today. Professor Chin reported at the time, "We're pretty sure this was dropped by *Tyrannosaurus*, but it can be tough to identify the poopetrator."

Dinosaur coprolites are forever popular, and the Poozeum, a museum devoted to coprolites and based in Florida, reports that it has identified an even larger *T. rex* coprolite, this monster measuring 26 inches (67 cm) long and named Barnum by its proud owners.

INFECTIONS AND PARASITES

Dinosaurs suffered all kinds of diseases, and many of these relate to their feeding and diet. In a 2009 paper, Ewan Wolff of the University of New Mexico and colleagues reported infections in the jaws of *T. rex* and relatives. Most of the jaw bones showed numbers of circular holes of similar size, and at first paleontologists thought they were bite marks. But the holes did not line up, and they occurred all over the jaw bones. On close inspection, the edges of the holes showed rings of continuing bone growth, so whatever had made the holes did not kill the animal and it continued growing, and its body was in some way fighting back.

The investigators then looked at veterinary literature on modern birds and found that holes such as these are produced by a protozoan parasite called *Trichomonas*, which gives modern birds throat ulcers and bone infections. The parasite can be passed on from infected prey or from other members of the species. So, if a *T. rex* ate some infected meat, or if they engaged in face-biting behavior, the infestation could pass around a herd. This idea has been challenged, however, by Bruce Rothschild and colleagues in 2022, who argue the puncture marks indicate biting by rival dinosaurs.

The parasite attacks were not all seen in the bone, but some are evident in the fossil poop.

PRIZE-WINNING POOP
This fantastic coprolite from *T. rex* is 17 inches (43 cm) long, and it contains the broken bones of a herbivorous dinosaur, evidence for what the huge predator had been eating.

FEEDING

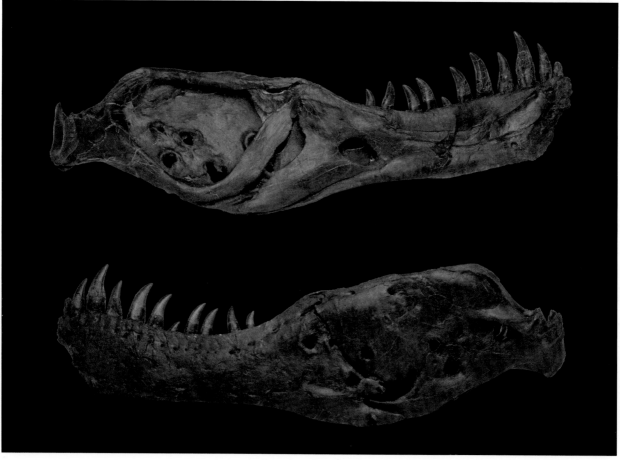

Today, wild animals suffer all kinds of gut parasites, worst of which is the tapeworm. This fiendish worm clings to the inside of the guts of modern animals, including birds and crocodilians, and feeds on their food as it passes into the gut.

Tapeworms do not need oxygen, and they are perfectly happy existing inside a dark internal space without any air. They can grow to huge length, and they breed by releasing ends of their body that pass out with poop and can then infect any other animal that noses around the poop. All dog-owners know that dogs pick up tapeworms all to easily, and the pets have to be regularly wormed to remove the parasites.

Science writer Carl Zimmer once asked whether dinosaurs had tapeworms. The answer is almost certainly yes. But then, how long would they have been? In humans, tapeworms can be up to 80 ft (25 m), curled up in the gut, making the host feel distinctly ill as the internal parasite consumes food. In the worst cases, the tapeworm can stay inside its host for up to thirty years, feeding away perfectly happily. At this rate, a dinosaur tapeworm might have been hundreds of feet (dozens of meters) long.

Researchers have identified intestinal worms in dinosaur poo. In one study, the investigators ground up a coprolite, treated it with acid to remove all the minerals, and laid out the organic remains under the microscope. They identified remains of a protozoan and eggs of three small worms, parasitic flukes, and roundworms similar to the kinds of parasites humans and animals get today from drinking polluted water.

All the gut parasites spread through their hosts by passing out in poop, which then either lies around and is touched or sniffed by other animals, or goes into river water, where the protozoans and eggs can survive for a long time until they are consumed by other animals. They hatch when they sense a warm, moist environment such as an animal's gut.

These weren't the only parasites to pester dinosaurs. We know there were mosquitoes and other bloodsucking insects in the Mesozoic, and these probably bothered dinosaurs just as much as horseflies drive large mammals mad with their biting and bloodsucking.

HOLEY JAW
The skull of a *T. rex*, top, shows large round holes through the lower jaw. In more detail, below, the jaw is seen from the inside and outside, and the holes can be seen to pass right through the bone, and have raised edges. This proves the holes were made while the animal was still alive and the bone began to grow again around the damaged areas. Investigators identified the holes as similar to those made in jaw bones of modern birds by parasites that give them throat ulcers and bone infections.

FORENSICS:
ISOTOPES AND DIET

We've just looked at all the kinds of fossil evidence for dinosaur diets, but modern chemistry offers another approach. Archeologists measure the status of certain common elements, such as carbon, oxygen, nitrogen, and lead, in ancient bones or teeth to work out where early humans might have come from. For example, lead enters drinking water from the rocks and soil, and different regions have characteristic values. What we drink leaves a chemical trace in bone, and it's been possible to identify, for example, that an ancient skeleton came from someone who was perhaps born in Germany, then lived in England. In other words, the values change as the person grew and this is laid down in the bones.

These common chemical elements occur in different forms called isotopes, and recognizing the relative amounts of different isotopes of, for example, carbon and nitrogen, can say a lot about ancient diets. In the graph below, more of the heavier isotopes of carbon indicate a switch from a diet of regular (so-called C3) plants, through omnivory ("eating everything"), carnivory, then feeding on seafood such as fish and clams, through to a diet of so-called C4 plants such as maize and sugarcane.

On the nitrogen isotope axis, the values become heavier as you go up through the food chain, for example from rabbit to fox, or salmon to seal. The plants and clams contain lighter isotopes of nitrogen. As these are consumed by the next step up the food chain (see page 69), the heavier

NITROGEN ISOTOPES AND DIET Measuring two versions of nitrogen values, corresponding to nitrogen-13 and nitrogen-15, gives a measure of the types of plants being eaten, from left to right, and the positions of animals in the food chain, from bottom to top. At the top of the food chain, the walrus eats the seal, which eats the salmon, which eats the smaller fish, which eats the clam, which eats the seaweed. Nitrogen ratios from fossil bones, then, can tell us the diets of extinct animals like dinosaurs. C3 and C4 plants use different kinds of photosynthesis; most plants are C3, and C4 plants include maize, sugarcane, and sorghum, commonly eaten by humans.

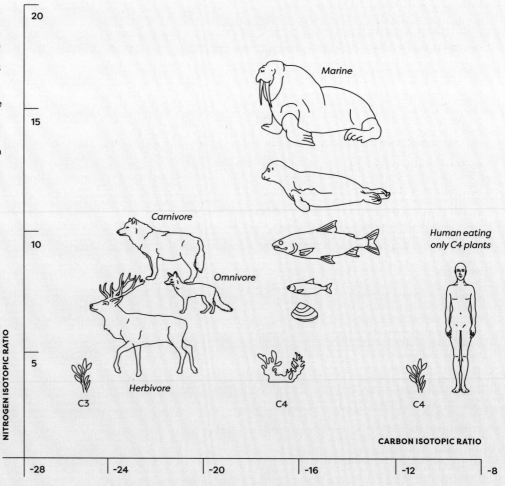

isotopes become more common, and so on up to the next step of carnivore.

In studying dinosaurs, such studies are beginning, but they can be tricky because the isotope differences are subtle and difficult to measure, and proportions of isotopes of carbon and oxygen may be affected by the climate as well as the food and water consumed. One study has shown a remarkable case of dietary shift in a dinosaur.

Limusaurus from the Early Cretaceous of China is known from skeletons of juveniles and adults, and it was noted that the young had teeth, but as the animal grew it shed its teeth and developed a beak. The researchers suggested this proved a dietary shift, from carnivory as a youngster, to omnivory or herbivory as an adult.

The researchers measured oxygen and carbon isotopes of a series of *Limusaurus* specimens and compared them with isotope measurements from other dinosaurs whose diet was known. In the graph below, the carnivores and herbivores occupied different areas. They found that *Limusaurus* babies sat in the top corner of the carnivore space, then more grown-up examples had hopped over to the herbivore space. A great example of how modern science can shed light on ancient mysteries.

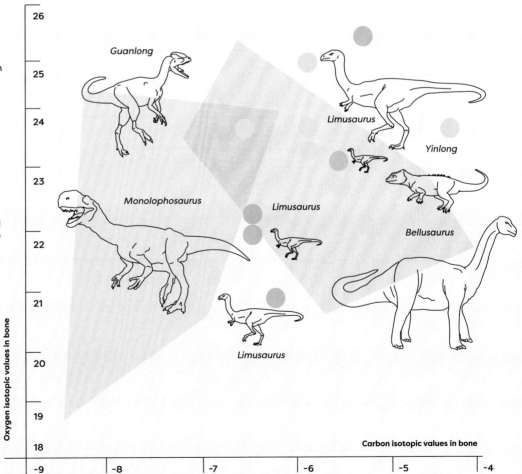

CHANGING DIET
How could we tell that a dinosaur switched from being a carnivore to a plant-eater later in life? It can be done by comparison of isotope values in the teeth. In modern animals, the balance of oxygen and carbon isotope ratios in teeth and bones varies according to their diets. At different ages, the small theropod *Limusaurus* shows different values, corresponding to changing diets, from youngest (green) to oldest (yellow). It shifts from being a carnivore when it was young (overlapping the pink shaded area of other theropod dinosaurs) to being a herbivore when it is older (shaded area overlapping other herbivorous dinosaurs).

- Carnivores
- Herbivores
- *Limusaurus* (babies and juveniles)
- *Limusaurus* (adults)

FEEDING

DINOSAUR HERBIVORY

The teeth of herbivorous dinosaurs give some clues about the plants they ate, but more reliable evidence comes from their dung and the fossil plants around them.

We know about the plants of the Mesozoic from their fossil leaves, stems, roots, bark, seeds, and spores. Many of these plants had edible leaves, including mosses, ferns, horsetail rushes, cycads, ginkgos, and evergreen conifers such as pine trees, redwoods, and their relatives. Some of these plant types seem exotic today. Cycads, for example, are woody plants that look like a tree trunk with a sprouting fern on top. Ginkgos, sometimes called maidenhair trees, still exist naturally in China today but were a much more diverse group in the Mesozoic.

One of the big events of the Cretaceous was the origin and spread of flowering plants, technically called angiosperms. Today there are over 300,000 species and they dominate the landscape everywhere. They include all the common trees, except the conifers, as well as shrubs and smaller plants, and all bear flowers. They include all the common food plants such as grains, cabbages, and root crops. Angiosperms progressively took over in Cretaceous landscapes, and even some rare grasses.

However, except for the case of the sauropod coprolites from India (see page 146), it seems dinosaurs rarely ate angiosperms, and they stuck with the ferns, horsetails, ginkgos, and conifers they had always eaten. This was perhaps not just that they were fussy eaters, but also that their jaw equipment and teeth had evolved to eat specific food types—such as the impressive dental battery of the hadrosaurs, designed to grind up conifer needles—that they could not switch to the new plants. Also, angiosperms commonly contain poisonous chemicals such as tannins, and herbivores have to evolve ways to cope with them.

SEED DISPERSAL IN DUNG

Plants are great survivors and many of them produce seeds covered with tough protective coats. When an animal chomps through a plant, swallowing leaves and seeds, for example in conifer cones, most of the plant material breaks down, providing nutrition to the herbivore, but often the seeds resist chemical action in the gut, and they pass out the other end in the herbivore poop.

Many dinosaur coprolites contain seeds that have been through the animal's gut, and they are now in a great situation to grow. They may be ripe and ready to sprout, and the dinosaur dung provides a perfect growing medium. Why would plants evolve this mechanism?

It's all about getting spread around. If a plant simply drops its seeds on the ground, like cones falling in a circle around a tall pine tree, they can grow where they fall, but then there is a cluster of hundreds of pines in a narrow spot. A herbivore can carry the seeds away, and this means the young plants might have a better chance to survive, not being overshadowed by the mother or father tree.

It's been estimated that, for example, the ceratopsian dinosaur *Triceratops* could have carried the seeds 2–3 miles (4–5 km) easily in the time it would take one meal to be digested and to pass through. This 10-ton dinosaur might have walked this distance in a day as it circled around looking for food. When it was migrating from one feeding area to another (see page 86), it might have covered 20 miles (35 km) in a day. This is a great way for a plant to spread rapidly over the landscape.

NICHE PARTITIONING

A key issue for herbivores is to avoid competition. In herds of plant-eaters out on the African plane, sometimes dozens of different species of deer and cattle are out there feeding side by side. In many cases they eat exactly the same grasses, but some may feed on taller grass, others on lower grass. Farmers know to put horses in a field first to

SEEDS IN GUT
The ornithopod *Isaberrysaura* from the Jurassic of Argentina shows preserved gut contents inside the rib cage. These consist of packed remains of seeds from cycads, a group of fern-like plants, which the dinsoaur had been feeding on just before it died.

Seeds

eat the long grass, then cattle or sheep to chomp at the lower stems. This is called niche partitioning, dividing up the food according to slightly different habits or methods. The "niche" of a species is its exact mode of life, including preferred food.

It was the same in the Mesozoic, and researchers had long wondered how so many species of herbivorous dinosaurs could live side by side. A classic example is in the Late Jurassic Morrison Formation, where seven or eight sauropods lived together. Their body shapes suggest one aspect of niche partitioning: tall *Brachiosaurus* almost certainly fed high in the trees, whereas low *Dicraeosaurus* fed closer to the ground. In fact, isotope studies (see page 150) confirmed this, showing that *Brachiosaurus* fed on conifer leaves, whereas *Dicraeosaurus* fed on ferns.

What if they were feeding at the same level. In a study of the skulls of *Camarasaurus* and *Diplodocus*, David Button of the University of Bristol, UK, showed they functioned differently. He applied FEA methods (see page 141) to the skulls and found that *Camarasaurus* had a more heavy-duty skull that could withstand greater forces. This shows it ate tougher food. Button argued that *Camarasaurus* and its relatives were generalized browsers that fed on hard and even woody material. On the other hand, *Diplodocus* and its relatives specialized on softer, but abrasive, plant materials such as horsetails and ferns.

Looking at the skulls, *Diplodocus* has a kind of goofy expression, with a bunch of long, narrow teeth right at the front of the jaws. This, and the stress patterns from the FEA study, suggest that it was a branch-stripper, feeding by grasping a leafy branch between its teeth and yanking back with the teeth clenched firmly shut. The leaves came off the branch, *Diplodocus* swallowed, then looked for another branch. *Camarasaurus* and relatives had heavier, shorter, spoon-shaped teeth along the whole length of their jaws, and they used a more regular feeding mode. It's likely they simply snipped and chomped bunches of branches and leaves, without separating leaves and woody material.

DIET SHIFTER
Studies of isotopes in the bones of *Limusaurus* (see page 151) show that juveniles were meat-eaters but adults were plant-eaters. Here, a juvenile swallows a lizard, while mom feeds on a cycad plant.

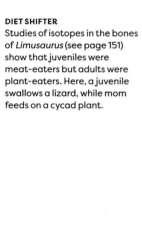

BROWSING LEVELS

Different sauropods fed at different heights. We can work this out two ways: one, by looking at the length of the neck; the other, by measuring carbon isotopes in the bones. *Brachiosaurus* has a long neck and is thought to have fed at the highest levels of trees, whereas short-necked *Dicraeosaurus* fed on bush-like plants. These two plant types contain different isotopes of carbon, and this can be measured from the bones of dinosaurs.

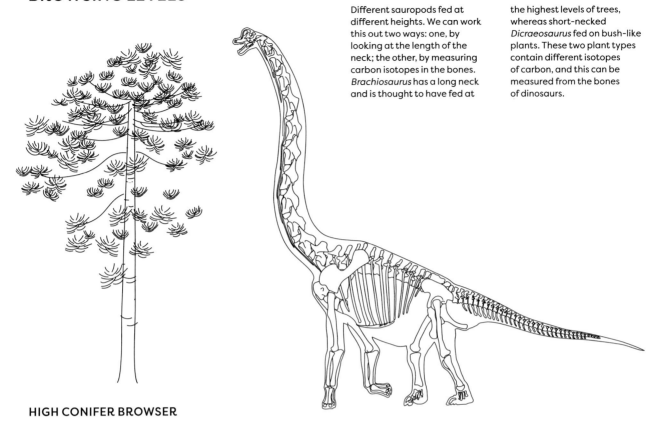

HIGH CONIFER BROWSER

Brachiosaurus
Higher $\partial^{13}C$ values
(carbon isotope values)

LOW FERN BROWSER

Dicraeosaurus
Lower $\partial^{13}C$ values
(carbon isotope values)

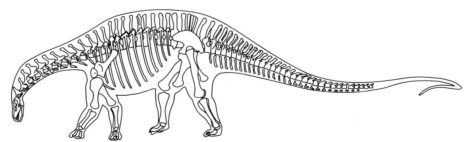

NEXT PAGE
Sauropods in the Morrison Formation of Wyoming show how they feed at different levels. *Diplodocus* feeds on low plants to the right, while *Brachiosaurus* on the left reach high in the treetops. *Camarasaurus*, in the foreground, ranges from low to mid-height in its search for edible leaves. These giants lived side by side with other dinosaurs such as *Stegosaurus* (deceased, skeleton in foreground) and the small ornithopods *Dryosaurus* (far right).

FEEDING

DINOSAUR CARNIVORY (AND OTHER DIETS)

In fact, dinosaurs were not simply herbivores or carnivores. Among the carnivores, there were some unusual specializations.

For example, some dinosaurs probably specialized in eating fish or insects. Also, some theropods switched from a diet of flesh to either a mixed diet of flesh and plants, called omnivory, or completely to herbivory.

The majority of theropods though were flesh-eaters, but their approaches differed. Of course, the smaller theropods fed on small prey such as small dinosaurs, but also lizards, mammals, frogs, and any other prey they could catch. Some would have grabbed a juicy cockroach or grasshopper when they could, as foxes and coyotes do today. The giant carnivores fed on larger prey, and *T. rex* would likely only eat an insect if it flew in his mouth or was already sitting on a piece of meat.

In a study of many features of the teeth and jaws of theropods, Joep Schaeffer, then

DIETS AND JAW MEASUREMENTS

The shape of the jaw identifies dinosaurian diets, and we can be sure of this by comparison with modern reptiles such as lizards and crocodiles. You simply take measurements of the jaws and teeth, run the multivariate statistical analysis, and see where the additional species falls on the plot to determine diet. Here we see a morphospace (representation of shape and form) based on taking dozens of jaw measurements, and the two axes show different aspects of jaw shape: toothless forms to the left; toothed forms to the right. Shallow jaws are shown at the bottom; deep jaws at the top. Then, the jaw shapes characteristic of herbivores and omnivores fall in one area, and small carnivores and large carnivores in others.

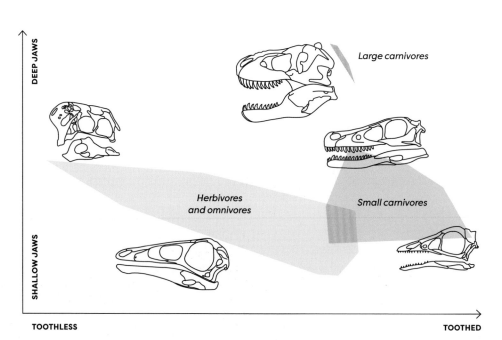

a student at the University of Bristol, UK, showed that these characteristics are quite distinct in the different categories. Note that he included only theropods here, so the herbivores and omnivores are theropods such as oviraptorosaurs and therizinosaurs, but their teeth and jaws were very different from those of their carnivorous relatives, and there is also a distinction between small and large carnivores.

HUNTING VS. SCAVENGING

There has long been debate about whether *T. rex* was a hunter or a scavenger. Scavengers are predators that mainly feed on dead meat. Hyenas and vultures can kill their prey, but they seem to prefer to home in on something killed by a lion or a jackal then dash in to grab some juicy strips of flesh.

Scavenging is a smart strategy because you don't lose a great deal of energy in chasing and killing the prey, and you can move on to another kill site if the locals aren't friendly. There are risks of course; for example, you might not find a fresh enough kill, or you might get killed by the lion or jackal, defending the results of all its hard work.

In the case of *T. rex*, the scavenging argument is that this was a huge animal, weighing over 5 tons, that needed plenty

WEIRD AND WONDERFUL
Therizinosaurs are some of the weirdest theropod dinosaurs. The Late Cretaceous *Therizinosaurus* was huge and had swordlike claws, 3 ft (1 m) long, on its hands but tiny teeth, so it was not a flesh-eater. It might have used its claws for raking together plant material, but it seems they might have been used mainly for display. Perhaps it flashed and clacked its claws to impress its mates.

FISH FOR SUPPER
Here, the spinosaurid *Baryonyx* flips a fish and prepares to swallow. Spinosaurids were strange theropods mainly from the Early Cretaceous, and their long-snouted, crocodile-like skull suggests they had a diet of fish. Some fish remains were found inside the rib cage of *Baryonyx*, so perhaps that was its last supper.

of food, and yet was probably a slow plodder. It would be wrong to imagine *T. rex* could run at the speed of an express train and leap and kill any herbivore it encountered. Others point to the silly little arms *T. rex* had, far too small to be of any use in grappling prey.

However, there are many arguments to say *T. rex* was a hunter. For example, the tooth trapped in the tail of a hadrosaur (see page 143) shows they did attack living animals. Also, there probably wasn't enough carrion, dead flesh, lying around (see page 28), so *T. rex* had to work for his supper.

SOLITARY OR PACK HUNTING?

Today we see large predators either hunting alone, like cats, or in packs, like hunting dogs. Sole hunting is fine if the predator is large and powerful and can bring down its prey. Pack hunting is great when there is a big size difference. So a pack of wolves can chase and harry a moose, which weighs ten times as much, for days until it tires, then the whole pack gets a substantial meal.

With dinosaurs it's hard to tell, because this is a kind of behavior that does not fossilize. But pack hunting has been suggested for the Early Cretaceous predator *Deinonychus*. It lived side by side with the ornithopod *Tenontosaurus*, weighing in at about 1 ton, compared to 160 lb (70 kg) for *Deinonychus*. One locality showed a single *Tenontosaurus* skeleton and three *Deinonychus* skeletons, and it was suggested that this was a hunting site, showing three raptors attacking a single large herbivore.

More likely, this was just a chance association, and most paleontologists would say we don't have good evidence that dinosaurs ever hunted in packs.

SPECIALIST THEROPODS: PLANTS, FISH, AND INSECTS

There were all kinds of plant-eating dinosaurs in the Late Cretaceous, such as hadrosaurs, ceratopsians, and ankylosaurs, but several groups of theropods, such as oviraptorosaurs, therizinosaurs, ornithomimids, and some troodontids, switched from flesh-eating to herbivory. We know this because they lost their teeth or evolved tiny teeth, or evolved new jaw shapes to bite in new ways.

What about fish-eating? When the skeleton of the spinosaurid *Baryonyx* was excavated in England, UK, paleontologists noticed it had fish scales inside its rib cage. Was this a specialist fish-eating theropod? Emily Rayfield studied its skull using FEA (see page 141), and she showed that these weird, long-snouted theropods might indeed have been fish-eaters; the jaws were built for speedy snapping, but not for tough fighting with larger prey.

Another strange diet is eating termites and ants. Zichuan Qin, a student in Bristol, UK, and Beijing, China, noticed that the alvarezsaurids, a group of theropods, became very small in the Late Cretaceous, and they just kept a single powerful finger on their short, punchy little arms. He compared these claws with modern animals and saw that they were shaped like digging claws in modern ant-eaters. So maybe these amazing little critters specialized on eating ants at a time when ants were becoming important parts of the Late Cretaceous ecosystem.

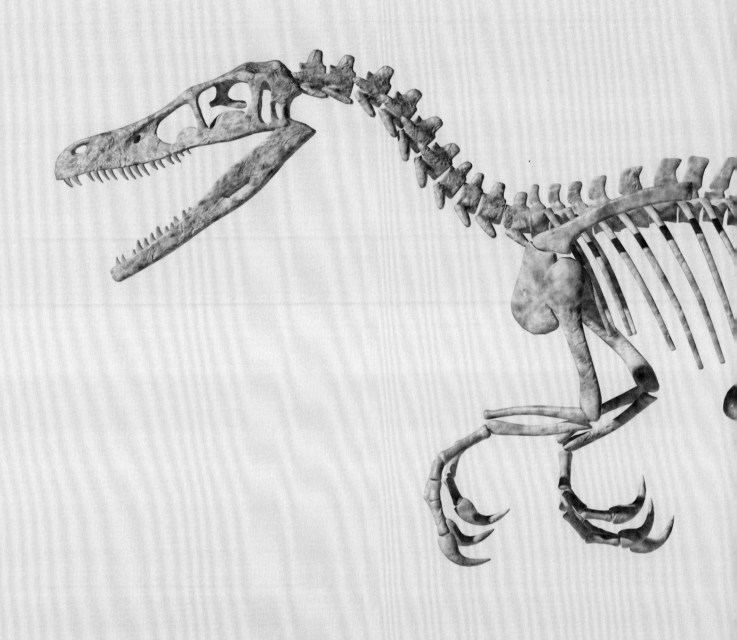

6

SOCIAL BEHAVIOR

COMMUNICATION AND INTERACTION

Paleontologists have identified a surprising amount of detail about dinosaur social behavior, from courtship and mating, through eggs and growth, to living in herds.

Social behavior includes all the interactions between members of a single species. Humans are very social, and we communicate all the time with members of our family and people we see in our daily life. Many animals do the same, and we assume dinosaurs did too. We've looked at some aspects of dinosaurian social behavior so far, including herbivores sharing feeding grounds and possible pack hunting by some carnivores (see page 161).

Social behavior is key during all phases of reproduction, and in the production and care of the young. There are also questions around how dinosaurs communicated with one another and whether they lived in herds, and if so, why.

In terms of mating and reproduction, of course dinosaurs did this but how? And were they more birdlike or reptilelike in their behavior? When we look at the living relatives of dinosaurs we think of birds laying eggs in nests, many of them showing shared parental care for the young, with the mother and father taking turns to sit on the eggs to keep them warm, and flying far and wide to gather food for the hungry babies. Reptiles, on the other hand, generally do not show much parental care. Dinosaurs laid eggs in nests on the ground, and there are many questions about whether they behaved more like birds or reptiles in caring, or not caring, for their young.

Before any eggs were fertilized and developed, the parents had to get together. We now know that many, or even all, dinosaurs engaged in courtship rituals. Like many birds today, evidently the male dinosaurs, at least the small feathery ones, hopped around, showing off their colorful, feathered tails and wings, hoping to gain mating access to one or more females. Many species of modern reptiles also show pre-mating rituals of one sort or another; male lizards or snakes, for example, show off their brightly colored patterns and crests.

All of these aspects of behavior would seem difficult to pin down in dinosaurs. How could we know their colors or patterns, or how they communicated, for example through sounds they made? Remarkably, we do now know something about dinosaur voices and colors. There are fossils that are beginning to tell us differences between males and females, and behavior of dinosaurs in herds.

These are exciting new areas of research, and there is huge potential for the young paleontologists of today to make some great discoveries.

FIGHT!
Dinosaurs probably had complex behavior just like modern animals. This doesn't mean they were always nice, and sometimes they had fights. Here, two male *Styracosaurus* fight over who is the strongest. Perhaps like modern deer and lions, the males controlled groups of females, and young males would sometimes have challenged the old, more powerful male.

SEXUAL SELECTION AND FEATHER COLOR

Dinosaurs seem to have shown differences between males and females, and engaged in pre-mating displays, just like modern peacocks—and we even know their feather colors!

One of the biggest surprises in recent years has been the discovery that many dinosaurs had brightly colored feathers, and even elaborate crests and tail fans (see page 171). Such structures could not be explained as essential for insulation or flight, the two other functions of feathers. An animal with a ginger head crest, or two stripy tail fans, or intensely striped wings is probably using them for display. Paleontologists had already seen that some dinosaurs had weird bony crests, frills, and horns that might have been used in display, but fancy feathers gave new life to this idea.

SEXUAL SELECTION

What's the point of pre-mating rituals? Why, for example, does a male peacock or pheasant have such a long, elaborate tail? The tails look amazing when displayed; the feathers upright, showing the pattern at its best, even accompanied by rustling noises as the birds vibrate the feathers. But in day-to-day life, surely such structures are a major disadvantage?

Charles Darwin (1809–82) worried about the fact that many display structures, especially in birds, might be so disadvantageous. He understood that in all other regards, plants and animals had evolved to be well adapted to their daily lives. Through time, competition between members of a species would tend to make the successful ones better at what they do, maybe smarter at finding food or quicker to escape predators. These are the sorts of adaptations that lie at the root of Darwinian evolution by natural selection.

Darwin reasoned that the male peacock with the smartest or largest set of tail feathers might have the greatest success in mating. Generally, male peacocks want to impress as many females as they can and mate with them, passing on their genes to the next generation. Darwin argued that the advantage to the male of carrying the crazy big tail all year round paid off during mating season because in the end, success in evolution is defined by the number of babies you leave behind. For the peacock, a great tail equals lots of babies.

But what about the females, the peahens? They have all the hard work in making the eggs and in parental care, and it's not great for them if the male mates and leaves. Darwin's argument is that the female uses the tail as a means to identify the strongest male, which ought to be the best father for her babies. This is what Darwin called sexual selection, to distinguish it from natural selection.

In nature, natural and sexual selection can be in tension: a bigger tail and more mating versus greater risk from predation. In some species of wild pheasants, the males live for only ten months, the females for twice as long. But if the short-lived male mates with two or three females and fathers five or six young, he is a great success in terms of evolution.

SEXUAL DIMORPHISM

Dimorphism means "two shapes," and the idea is to identify cases where dinosaurs show evidence of differences between females and males. In birds, as we have seen, there are many cases where the males have brightly colored feathers and the females do not. The female pheasant and peahen are drab brown in color. In other birds, the male and female may have different songs. In mammals, males are often much larger than females, so they can fight other males, often using specialized weapons such as tusks (baboons, pigs), antlers (deer), or horns (cattle, sheep).

MALE VS. FEMALE

Male and female *Stegosaurus* may have differed in the shapes of the plates down their backs, with females showing higher, narrower plates, and males having larger, more rounded plates. The males may also have had a neck wattle, loose skin beneath the neck.

SOCIAL BEHAVIOR

Identifying sexual dimorphism in dinosaurs has been difficult. Examples have included supposed differences in the head crests of hadrosaurs, horns on ceratopsian skulls, and arrangements of armor plates and spines in stegosaurs. Closer study has cast doubts on some such examples when, for example, it is noted that the supposed males lived at a different time or in a different part of the world than the supposed females.

It has been suggested recently that *Stegosaurus* showed dimorphism in the shapes of its back plates—tall and narrow in one form, more rounded in the other. Also, in the ornithopod *Maiasaura*, famed for detailed information about its nesting and babies (see pages 180–1), one sex is 45 percent larger than the other when they were fully adult. But in these cases, we can't tell which is male and which is female, and it would be wrong to assume that large dinosaurs are always males. In many reptiles today, such as snakes and turtles, females can be larger than males as they have to carry large numbers of eggs.

SEXUAL SELECTION IN *CONFUCIUSORNIS*

One of the first fossils to be studied for sexual selection was the early bird *Confuciusornis* from the Early Cretaceous of China. In a famous specimen, a pair of birds are preserved in a single slab of rock, and one of them—the bigger one—has a pair of long thin tail feathers, each ending in a kind of expanded banner, whereas the smaller example has no such tail feathers. Can paleontologists demonstrate that this is direct evidence for sexual selection in one of the very first birds?

In a 2008 study, Luis Chiappe from the Natural History Museum of Los Angeles could not confirm anything so simple. He measured a hundred specimens and found there were two size groups, one around 10½ oz (300 g), the other around 17½ oz (500 g). Could these groups be males and females? The two size sets could even represent two different ages, maybe teenagers and adults, born a year or two apart. Also, they noted that birds large and small sometimes had the long tail feathers, sometimes not.

In a study in 2013, Anusuya Chinsamy-Turan from the South African Museum identified definite evidence that one of the short-tailed specimens was female. It showed so-called medullary bone in the arm and leg bones. This special kind of bone is formed in modern female birds when they are ready to lay their eggs. They extract large amounts of calcium into their bloodstream to build the eggshell, and this leaves a particular irregular, spongy kind of bone behind, the medullary bone.

LEFT AND ABOVE
Just as many birds today show definite differences between males and females, this was true of the early bird *Confuciusornis* from the Early Cretaceous of China. Males and females had a colorful mix of gray and brown feathers, with spots and spangles, but the males had a pair of long, dark-colored tail feathers. Just like peacocks today, they probably jumped high or posed on tree branches. They could rustle those long feathers to show the females how great they were.

Maybe the long tail feathers grew each year in time for the mating season and were then shed. Then, you would see the long tail feathers in birds of different sizes, and whether these feathers were present or not could be explained by the season of the year. But did only males sprout them, or were they present in males and females? Chinsamy had certainly identified a female bird with short tail feathers, but did the bird grow long tail feathers at other times of the year?

SEXUAL DISPLAY

There are not many studies yet showing how dinosaurs might have used their feathers and colors in display, but most paleontologists are convinced that they did. For example, it's likely such dinosaurs as *Sinosauropteryx* and *Anchiornis*, whose colors were first to be determined (see pages 172–3), used their colored wings, tails, and head crests in pre-mating display.

Sinosauropteryx had a neatly striped ginger and white tail, which would have made an excellent display structure. Likewise, *Anchiornis* had a ginger head crest and sharply defined black and white stripes on its wings—it's easy to imagine that this dinosaur in display would have ducked and raised its wings and tail, just like a lyrebird today, to show off the size and grandeur of its color patterns.

Many specimens of the oviraptorosaur theropod *Caudipteryx* from the Early Cretaceous of China show a fan of about twenty pennaceous feathers at the tip of the shortish tail. Individual feathers are 6 inches (15 cm) long, and they carry cross bars of black and gray colors. In life it is likely that *Caudipteryx* individuals could expand the feather fan because each feather would have had a muscle at the base, and they might even have been able to rustle the feathers against one another, making a whirring noise, just as peacocks do today.

LEFT AND BELOW
It wasn't just the birds: dinosaurs also probably showed sexual displays. Below, a *Sinosauropteryx* waves its ginger-and-white stripy tail. Perhaps whole groups of the little dinosaurs waved their tails to show off who had the longest or brightest patterns. Over on the left, an *Anchiornis* dances on tippy toes, and expands all the feathers of its tail and wings to show off a black-and-white barred display.

SOCIAL BEHAVIOR

FORENSICS: TELLING THE COLOR OF FOSSIL FEATHERS

What colors were dinosaur feathers? As we saw earlier (see page 60), many theropod dinosaurs from China had rich feather coverings, sometimes consisting of many different feather types. The Jurassic ornithopod dinosaur *Kulindadromeus* from Russia had a remarkable mix of feathers and scales of different styles over all regions of the body, some evidently for insulation, others perhaps for temperature control or protection.

In many cases, the fossil feathers seem to show patterns, but the original colors cannot be seen directly in the fossils as preserved. For example, in the *Confuciusornis* tail feathers, they are just a different shade of brown, yet the original color pigments have chemically degraded, and the original bright colors seem to have been lost.

Then, a method to identify feather colors in fossils was suggested in 2008 by Jakob Vinther, then a graduate student at Yale University. He showed that fossil feathers contained melanosomes, specialized structures that contain the pigment melanin, which could only be seen under the scanning electron microscope. These are about 1 micron long, one millionth of a meter.

Melanin is a common pigment in nearly all living things, from plants and mushrooms to animals, and it occurs inside the body, around the guts and brain for example, and in the skin. One kind of melanin called eumelanin gives black, brown, and blond colors, and another called pheomelanin gives ginger colors. So, a ginger-haired person has pheomelanin in their hair, as does a red squirrel.

All hair colors in mammals are produced by melanin, as are many colors in bird feathers, and in these cases the melanin enters the hair or feather from its site of production in the skin and is contained in a tiny structure called a melanosome. Melanosomes with eumelanin are sausage-shaped, those with pheomelanin are ball-shaped.

So, two teams of researchers applied these methods to feathers of *Sinosauropteryx* and *Anchiornis* and identified their original colors, in 2010 studies. Since then, dozens of other early birds and dinosaurs have been studied and their colors and patterns reconstructed using this method. The new method has revolutionized how paleoartists reconstruct many dinosaurs (see page 51).

However, we can't tell the colors of all dinosaurs. The method requires preserved feathers or skin, and most dinosaur fossils do not show these materials. Also, we don't know whether early birds and dinosaurs had other pigments found in modern birds, such as carotenes (giving yellow and red colors) and porphyrins (giving green and purple colors). These do not leave a physical trace in fossils, and researchers are looking at chemical methods to try to identify them.

MELANOSOMES AND FEATHER COLOR

Fossil feathers, like modern feathers, contain melanosomes, tiny capsules that contain the pigment melanin. The shapes of melanosomes give us clues about the original color, as seen in feathers of modern birds. Thick rods are found in birds with weak iridescent coloring such as in the brown-headed cowbird (top row, left), while all other types produce brilliant iridescence, with the iridescence and its range largely tied to the thickness of distinct melanin layers in the structures, as seen in the Nicobar pigeon (second left), the trogon, sunbird, and hummingbird (right).

BARBULE CROSS SECTIONS

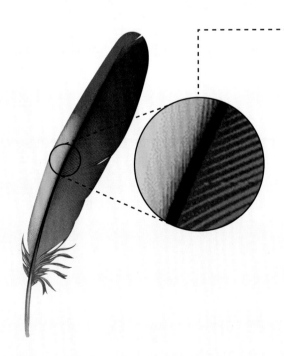

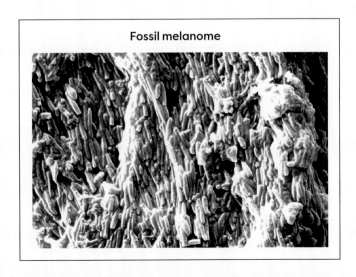

Fossil melanome

SOCIAL BEHAVIOR

COURTSHIP AND MATING BEHAVIOR

We know dinosaurs courted, and there are even suggestions they did it communally, but the only hint so far of how they mated comes from an amazing specimen from China.

To date, no fossil has been found of a pair of dinosaurs locked in embrace, so speculations about mating positions remain just that. But a new fossil from China shows the cloaca—the combined opening for passing urine and feces, as well as for exchange of sperm to fertilize the egg—and reveals some important clues regarding mating behavior. There are also some tracks from Colorado that may show a theropod courtship ground.

PSITTACOSAURUS CLOACA

In a 2021 paper, Jakob Vinther from the University of Bristol, UK, reported a detailed study of the cloaca of the ceratopsian dinosaur *Psittacosaurus* from the Early Cretaceous of China. The preserved specimen does not indicate whether this was a male or female, but it shows evidence of dark coloration around the cloaca in a series of scales of different sizes.

The *Psittacosaurus* cloaca is V-shaped and has broad lips on either side, to some extent like modern crocodilians. In the fossil, there is a fragment of coprolite partly on the way out. The dark coloring and thick lips around the cloaca are interpreted to show that this structure was displayed during the mating season, to form an attraction between male and female, and the animals might have released pungent scents through it, as in modern crocodilians.

THEROPOD COURTSHIP

One of the most unexpected claims from Martin Lockley's studies of the mega trackway sites of the American West (see page 84) was evidence for a display ground where theropods danced and showed off to one another. The evidence at first does not perhaps look very dramatic—large, scraped hollows in the ground, each 6 ft (2 m) across, but they do seem to be linked with tracks and footprints of theropod dinosaurs.

The paleontologists compared their interpretation to what certain birds do today. Peacocks, red grouse, ostriches, oystercatchers, and many other ground-nesting birds engage in what ornithologists call "lekking behavior." The male bird builds a display ground by clearing twigs and stones, and he prepares to dance when a female bird comes by. Sometimes, multiple male birds prepare the lek, or lekking ground, and each bird attempts to impress the females as much as they can by dancing, squawking, and scraping holes in the ground.

This is what Lockley and colleagues argue they had found in the Cretaceous Dakota Sandstone of Colorado. At four sites, they found the hollows in the ground, and could show that these had been made by backward scraping by the theropod dinosaurs using either their left or right foot. At one site, three large scrapes occur side by side, measuring over 18 ft (5 m). At the largest site, they found about sixty separate scrape structures, suggesting this was a location where multiple males were showing off, flashing their decorative feathers (see pages 170–1) and perhaps vocalizing.

THEROPOD MATING DANCE
Some footprint sites show strange sets of scratch marks and hollows. These have been interpreted as relics of an ancient mating display site, where large theropods danced, maybe in pairs, as some birds do, moving their heads up and down, and kicking up dirt. The reconstruction here is based on *Acrocanthosaurus* from the Early Cretaceous of North America.

EGGS AND PARENTAL CARE

There are many examples of dinosaur eggs, babies, and regular nesting grounds, but evidence for parental care is limited, and the babies soon had to fend for themselves.

Birds and reptiles today lay eggs, and it's no surprise that dinosaurs did as well. Biologists call egg-laying "ovipary," to distinguish it from cases where animals give birth to live young, which is called "vivipary." Mammals such as humans, cats, and dogs do not lay eggs, but the eggs develop entirely within the mother, and the baby is born directly.

Biologists have discovered that there are all kinds of variations on ovipary and vivipary. First, many reptiles, especially some lizards and snakes, lay soft-shelled eggs that have a tough, papery protective cover but no calcite mineralization. We are so used to the hard, white shells of bird eggs, as well as turtle eggs and eggs of some lizards, that we think this is the only kind or the main type of egg.

Recent studies of lizards and snakes have shown that things get complicated; some lay eggs, some bear live young, and it seems that they can switch easily from one mode to the other. There are several examples of closely related groups of lizards where some species lay eggs and their close relatives do not. It's obviously been possible for them to switch mode relatively easily and in both directions, from ovipary to vivipary, and even from vivipary to ovipary.

Why switch? The usual argument is that vivipary is good for survival of the babies because the mother effectively keeps them safe inside her until the right moment for birth, for example when the weather is good or food is plentiful. The related lizard that lays eggs might release them a week or so earlier when the air is cold, or plants or bugs are rare. Either way, the parents offer their babies little or no help, so their survival is uncertain.

DINOSAUR EGGS

So, what did dinosaur eggs look like? Some were shaped like a hen's egg, some were perfectly spherical, or ball-shaped, and others were long and sausage-shaped. They varied in size, depending on the species, and most are in the range between a hen's egg and a soccerball in size and shape. There were some whoppers, though, including some from the Late Cretaceous of China that are more than 2 ft (60 cm) long and about 8 inches (20 cm) across.

DINOSAURS BUILT NESTS
Here a *Heyuannia* parent shuffles about before sitting down to incubate the eggs. Dinosaur nests were shallow circles in the dirt, and the eggs here form one or two circles. The parent had to shuffle the eggs to the sides so they were not crushed, and spread its wings to warm the eggs.

New studies of the chemistry of dinosaur eggshells show that dinosaur eggs were brightly colored. Just as bird eggs can be white, yellow, blue, or brown, and uniform in color or speckled, it seems dinosaurs showed a similar range of colors. As Jasmina Wiemann of Yale University explained, dinosaurs could see color, and pure white eggs would have stood out clearly when they had been laid in ground nests. The researchers showed that the eggs of *Deinonychus* were blue-green, the troodontids had eggshells of blue-green, beige, or white, and the eggs of the oviraptorosaur *Heyuannia* were deep blue-green. These are the colors we see, but dinosaurs probably saw the colors differently, and it's likely they had all evolved as camouflage to protect the eggs from hungry hunters.

Were dinosaur eggs all hard-shelled? It seems that they were not. A 2021 study of all the dinosaur eggs showed that some of them laid soft-shelled eggs. When the investigators reconstructed the evolution of dinosaur eggs, they determined that the first dinosaurs probably laid soft-shelled eggs, too.

This remarkable conclusion makes sense of a conundrum that had puzzled paleontologists for a long time, which is that we know terrestrial eggs must have originated when reptiles arose in the mid-Carboniferous, and yet the oldest eggs found as fossils are Early Jurassic in age, 120 million years later. Where did all the missing eggs go? Were they just not preserved? That seems unlikely, because hard-shelled eggs preserve well normally. So, maybe they didn't exist and early reptiles all laid soft-shelled eggs, or even gave birth to live young.

EGG THIEF OR GOOD MOTHER?

One of the most controversial cases in dinosaur behavior concerns *Oviraptor*, a slender theropod from the Late Cretaceous of Mongolia. When the first skeletons were found in the 1920s, they were close to nests full of eggs, and paleontologists at the time thought the toothless, beaked predator was an egg-eater, so they named it *Oviraptor*, meaning "egg thief." Specimens and scenes were shown at the American Museum of Natural History (AMNH) in New York, with the evil *Oviraptor* menacing the small ceratopsian herbivore *Protoceratops,* which was trying to protect its nest from the predator.

In 1995, a new generation of AMNH paleontologists went to Mongolia and found an amazing skeleton of the oviraptorosaur *Citipati* actually sitting on the supposed *Protoceratops* nest. They X-rayed the eggs and found embryo oviraptorosaurs inside, not ceratopsians. Case proved!

EVIDENCE OF PARENTAL CARE
Here, the skeleton of an adult *Citipati* is shown with its long arms wrapping around the eggs. There are five long eggs on the right, and the hand with its huge claws holds them. The dinosaur's leg bones are in the center, showing how it must have brought its legs together and sat down carefully, so as not to break any of the eggs!

So how could a 9 ft (2.9 m) dinosaur weighing 175 lb (80 kg) incubate its eggs? Surely it would smash them to bits when it sat down! In fact, the fossil shows how it was done. The mother (or father, we don't know which) placed her long feet in the middle of the ring of fifteen to twenty eggs, lowered her bottom slightly, and shuffled her tail from left to right, wafting the eggs out to the sides. Then she lowered her body into the narrow space between the two egg rows and embraced them under her long feathery arms. She was incubating her eggs just as birds do today, using her fluffy feathers to keep them warm.

NESTS AND NESTING GROUNDS

So far as we know, dinosaurs all made their nests on the ground, scratching and digging with their hands and feet to shift the dirt and make a hollow space. Whether any dinosaurs made nests in trees we do not know, nor even whether *Archaeopteryx* and the first birds did so or not. Today, many birds also nest on the ground, so tree nesting might have come much later.

The dinosaur mother then squatted over her nest scrape and deposited her eggs, usually in a definite pattern; a circle as in *Oviraptor*, or a double row as in many sauropods.

A recent discovery is that dinosaur eggshells were highly porous, just like crocodile eggs, which means their parents probably covered their nests in vegetation, both for protection and to provide some warmth as the plants decayed—a kind of compost heating system. This is what crocodilians do today, unlike birds that tend to lay their eggs in open nests and incubate them. Bird eggs are less porous. This makes sense—maybe slender dinosaurs such as *Oviraptor* could incubate their eggs, but a *Brontosaurus* parent would surely smash them if they tried!

Dinosaur mothers often made their nests in nesting grounds, areas where dozens or hundreds of other members of the same species congregated. A remarkable site in the Late Cretaceous of Argentina called Auca Mahuevo shows hundreds of nests of a sauropod close to *Saltasaurus*. In each scooped nest were fifteen to forty spherical eggs, each 6 inches (15 cm) across, and there were six egg-bearing layers. This suggests the mothers came back to the same site year after year.

This is an example of what ornithologists call nest-site fidelity, where the birds favor the same spot each year for egg laying, and it can be a successful strategy. Just as seabirds today nest in high cliffs to protect their babies, perhaps the dinosaur nesting site was safe from predators or had suitable soil for digging.

RAISING BABIES
Maiasaura from the Late Cretaceous of Montana has been found close to nests with eggs and babies. Did the mom and pop stick around while the eggs hatched and help feed their babies for the first weeks or months of their lives?

DINOSAUR BEHAVIOR

PARENTAL CARE OR NOT?

In the 1970s and 1980s, Jack Horner of the Museum of the Rockies, Montana, excavated similar nesting grounds with nest site fidelity in Montana and found a great deal of evidence for parental care. He and colleagues named the dinosaur, a Late Cretaceous hadrosaur, *Maiasaura*, meaning "good mother reptile." We've already seen that some dinosaurs incubated their eggs, but there was some evidence that *Maiasaura* parents also brought food to their hatchlings to help them for the first weeks after they left the egg.

What would we expect? If we consider living relatives of dinosaurs, it's well known that the bird parents invest a great deal of effort in bringing up their babies, which mostly hatch from the eggs naked, with eyes still shut, and helpless. The parents spend weeks or months foraging for food to feed their babies and eventually supervise their first efforts to fly.

In the 1800s, early naturalists watched as crocodile mothers seemed to be eating their babies. This contributed to the idea that crocodiles and alligators are nasty beasts. However, in these cases, the naturalists were wrong. The mother crocodile had stayed around her nest for up to three months, protecting the eggs, and was carrying her babies in her great toothy mouth down to the water, so they could swim and hunt for food.

HATCHING
The tiny skeleton of a baby *Maiasaura* struggles to hatch out of its eggshell. This is not an actual fossil but a reconstruction based on fossils. The baby had a hardened toothlike structure on its snout to help it break through the hard eggshell from inside.

SOCIAL BEHAVIOR

BABIES ON THE MOVE
Many dinosaurs were tiny when they hatched out, and some, such as *Psittacosaurus* from the Early Cretaceous of China, hung out with the other babies until they were maybe two or three years old. These baby creches might have fended for themselves, or sometimes an adult might have helped them at times of danger.

CRECHES

Dinosaur babies started out small, and this meant they had to grow up fast in their early years (see page 188). In many cases, young dinosaurs hung out together in small groups. The best examples are from *Psittacosaurus*, an Early Cretaceous dinosaur from China. Paleontologists were puzzled when they found hundreds of examples of young dinosaurs in small groups, maybe five to ten little beasts, usually without any adults, and sometimes all facing in the same direction.

The fossils come from a village called Lujiatun, and they are preserved in volcanic ash. It seems that volcanoes were erupting from time to time, and the small dinosaurs were caught in the air-fall ash. This is hot ash thrown out of the volcano that falls thickly over the landscape. It was so hot it fried the poor little dinosaurs and buried them at the same time. Maybe they were all facing the same way in their little clusters because they were running away from the threat.

But what were these little baby dino pods? Zhao Qi, a paleontologist in Bristol, UK, and Beijing, China, looked at the ages of the dinosaurs in one clutch, and found that five of them were two years old, and one was three. He suggested this was big sister (or brother) hanging out with younger siblings. Maybe they stuck together for safety.

These youngsters were herbivores, only 1 ft (30 cm) long, feeding on ferns and seed ferns, and they could hide in the vegetation from larger dinosaurs. Maybe their parents had helped them as tiny hatchlings, or perhaps not, but by the age of one or two they were on their own.

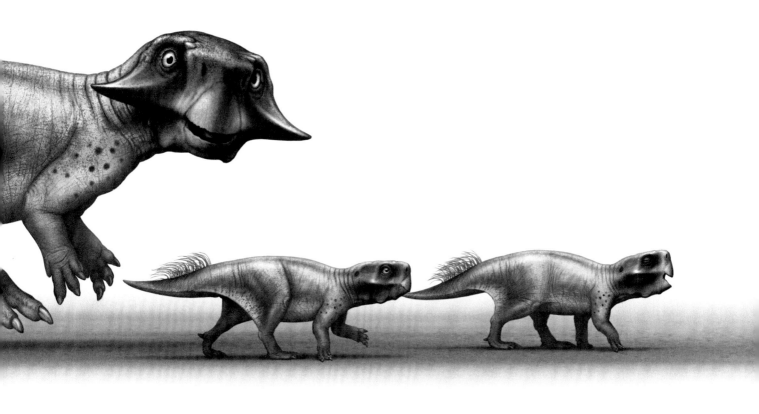

A CLUTCH OF YOUNGSTERS
Here there are six young *Psittacosaurus* in a single block comprising solidified volcanic ash. It seems the babies were running away from the volcanic eruption but were buried in hot ash and died, just as we see them.

GROWING UP

Baby dinosaurs were tiny and had to grow fast; paleontologists work out growth rates from comparisons of age versus body mass, and some are huge, adding around 5 tons per year.

Babies can be all different sizes and size often depends on what is expected of them. Human babies seem tiny and take many years to reach adult size. But by two years old, a human child is already half its adult height. Baby cattle and elephants are born large, with long, strong legs, and they are soon up on their feet, running beside the herd one or two days after birth. Some baby birds grow quickly to be almost as big as their parents, demanding more and more food as they do so.

But dinosaur babies were usually tiny because of limits on egg size. This meant they either took years to grow to adult size, or they grew very fast. New research shows that the latter was the case.

THE CLAMOR OF NEW BABIES
Here, some baby *Saltasaurus* sauropods are hatching at the same time. Each nest of eggs was laid by a different mother, but they laid their eggs in huge nesting fields where maybe hundreds of dinosaurs got together at egg-laying time. The babies are ready to start finding plant food, but they are tiny in size compared to their parents.

WHY WEREN'T DINOSAUR EGGS HUGE?

The size of eggs is roughly in proportion to the size of the mother—but not always. Mostly, large birds lay large eggs, but there's a limit to the size because of the thickness of the shell. In bird and dinosaur eggs, the hard shell is made of the mineral calcite, which is arranged in crystals that span the depth of the shell, and the eggs from different bird and dinosaur groups can be recognized by the crystal pattern inside the eggshell.

The key point is that in hard-shelled eggs the thickness of the eggshell is in proportion to the size of the egg: small eggs have thin shells; large eggs have thick shells. For example, the ostrich egg is 6 inches (15 cm) long, and its eggshell is 0.13 inches (3.5 mm) thick, like a thin slice of toast. In a 2 ft (60 cm) egg, the eggshell is three times as thick, and this is essential so the shell is strong enough to survive rolling around. If huge dinosaurs laid eggs in proportion to their body sizes, some might have been as large as a small car, and the hard shell would be 6 inches (15 cm) thick.

The result would be that the baby could not get out. When a baby bird, turtle, or dinosaur is ready to hatch, it uses a small toothlike structure on its nose to bust out. It can push and shove from the inside, and eventually make a crack and force its way out. If the eggshell were too thick, the baby would be stuck. So, this is one reason why dinosaurs laid relatively small eggs. The other reason is that it saved the mother a great deal of energy, part of a strategy for survival at such a huge size (see page 97).

SMALL BABIES

In the early dinosaurs, which were about human-sized, the babies were not so tiny. But as dinosaurs grew larger, their eggs stayed about the same size, like an American football. This meant the babies of a large sauropod were tiny in proportion to their parents, which

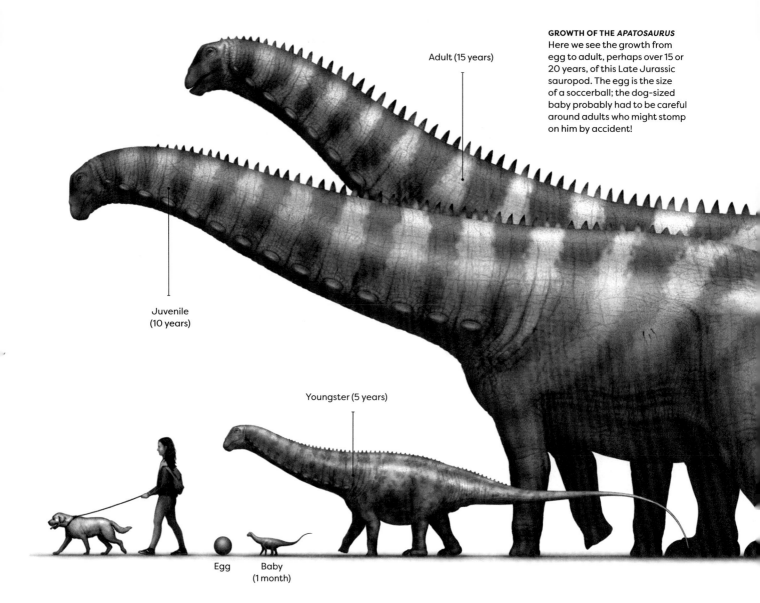

GROWTH OF THE APATOSAURUS
Here we see the growth from egg to adult, perhaps over 15 or 20 years, of this Late Jurassic sauropod. The egg is the size of a soccerball; the dog-sized baby probably had to be careful around adults who might stomp on him by accident!

in turn meant two things: they survived best if they kept well clear of the huge feet of their mom and dad; and they had to grow amazingly fast in their first year or two of life to make themselves safe.

A baby *Saltasaurus* hatched out at 1 ft (30 cm) long in comparison to its 28 ft (8.5 m) long parents. So, it had a lot of growing to do. A big question then is how much time it took for a large dinosaur to grow from tiny baby to large adult. One estimate was based on growth rates of cold-blooded crocodilians, and this suggested that sauropods might have taken a hundred years to reach adult size. However, the risk behind such slow growth is that most animals would die from predation or accidents before they ever reached a large enough size to breed. Not good, from an evolutionary point of view!

In fact, current evidence shows that many large dinosaurs reached adult size by age fifteen or twenty, even though this implies phenomenally fast rates of growth. The evidence comes from new ways of estimating age and body mass, which allows paleontologists to plot a growth curve (see page 189).

SKULL OF A BABY GIANT
The skull of a hatchling sauropod, *Saltasaurus*, from the Late Cretaceous of Argentina, is tiny, only 1 inch (2.5 cm) long. It was found close to a hunk of eggshell (top image) and in detail (below) shows how it lacks teeth. Like a human baby, it has large eyes and a short nose. It would later take maybe 20 years to achieve full adult size, when the skull would be 3 ft (1 m) long.

CUTE BABIES

Baby dinosaurs also looked very cute, or they would have done to us at least. In a recent study of a tiny hatchling saltasaurid skull, it could be seen that the face had a short snout and big eyes. This little critter would have looked extremely appealing when it hatched out, wobbling on its short legs and peeping with wide eyes. But the sauropod parents probably didn't notice any of this.

The big head, eyes, and knees have practical functions. Baby dinosaurs (and humans) have big heads and big eyes because the brain and eye in the baby are nearer adult size than the rest of the body to aid their functions. Also, elbows and knees in babies are big because they need to function well through life, and it's easier if they don't change size much as the baby "grows into them."

The study of the saltasaurid hatchling showed another thing. The little beast had an egg tooth still attached to its snout, ready to use to break out of its shell.

FORENSICS:
MEASURING GROWTH RATE

It's now possible to draw a detailed growth curve for certain dinosaurs. For example, we know that *T. rex* grew to adult size in twenty-five years, reaching its full size of 5–6 tons, and during its maximum growth spurt it was putting on 1,690 lb (767 kg) per year, which is nearly 5 lb (2.3 kg) per day. This matches the fastest growth rates of elephants, ostriches, and other warm-blooded animals.

In their growth, dinosaurs were more like birds than slow-growing, cold-blooded crocodilians, and this confirms other evidence for endothermy, such as their bone structure and possession of insulating feathers (see page 58).

To construct a growth curve, paleontologist Greg Erickson of Florida State University needed two measurements: the age and body mass of each skeleton. The age comes from growth rings in bone, and the body size from estimates based on body length, leg length, or diameter of the thigh bone (the femur). In fact, the best indicator of body mass (weight) is the diameter of the leg as represented by the femur, because the legs are proportional in width to the body mass they have to carry around. Large, heavy animals such as elephants and bison have thick legs, whereas lighter-weight relatives such as small deer and goats have skinny little legs.

Age can be estimated from growth lines in the bone. As we have seen, dinosaur bone is often beautifully preserved (see page 57), and in thin section under the microscope all the detail can be seen as if it were fresh bone. A key feature is the growth lines, technically called lines of arrested growth (LAGs). As any dinosaur grew in size, it had fast growing seasons and slow growing seasons. When food was abundant, perhaps in summer, the animal put on weight fast and the bone is open and pale in color, whereas when food was limited, perhaps in winter, growth slowed down, and there is a darker, more compact strip. Bone adds on the outside, as in growth of a tree, so you can track all sorts of aspects of a dinosaur's life history by counting LAGs across a leg bone or a rib.

Erickson was able to show that tyrannosaurs of different sizes reached adult, breeding size at different times, from fifteen to twenty-five years, depending on size. In fact, they all grew slowly up to age five, then rapidly from five to fifteen, sometimes putting on more than half a ton a year. The sauropod *Apatosaurus* grew even faster. It took fifteen to twenty years to reach its adult size of 30 tons, putting on more than 5 tons per year, probably the fastest growth rate of any animal, ever.

FEATURE DEVELOPMENT
This series of *Triceratops* heads shows how size increases from baby, at the bottom, to adult, at the top. But not only size changes, also the key features do. Notice how the neck frill becomes larger and frillier, and the horns start off tiny in the baby then become long and dangerous in the adult.

TYRANNOSAURS GROW UP
The graph shows how body weight increases with age, and we've used *Tyrannosaurus* as a visual example. Age is estimated from growth rings in the bone; body mass from bone sizes and overall skeleton size. Each genus of tyrannosaur reaches a different final adult size, ranging from 1 to 6 tons, and they become adults at anything from 15 to 20 years.

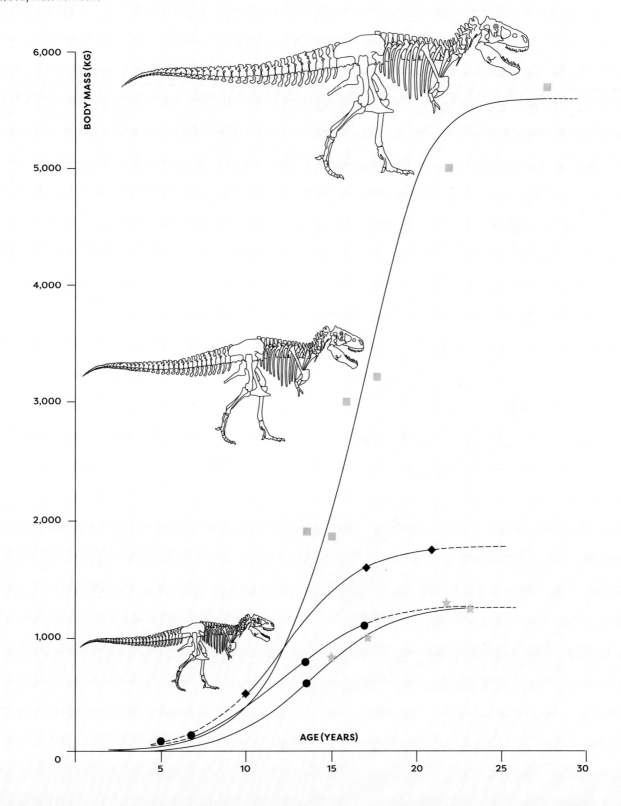

SOCIAL BEHAVIOR

COMMUNICATION AND LIVING IN HERDS

Animals communicate in many ways, especially if they live in large groups, and it is likely dinosaurs used their voices and the shapes of crests and horns to communicate.

We think of communication as speaking. This is because we are social animals who spend a lot of time passing information and explaining our thoughts and motivations to one another. We do this through language, and there are thousands of human languages, most of them highly complex.

Animals and plants communicate in other ways, including through their appearance, color, and displays. Birds communicate by their squawks and songs, as well as through visual displays such as feather colors and patterns, and especially special dances and tail displays.

Crocodilians have a variety of vocal communications including threat, distress, and "booming." The booms are below the normal human hearing range but can be heard for some distance through water. The dominant male in a group goes "boom, boom, boom," presumably saying, "Clear off, other males! Females, you should mate with me. I'm great!"

In the case of dinosaurs, we can see their crests, horns, feathers, and even their colors. We can even reconstruct the sounds that some hadrosaurs made (see page 194). So, it's likely they were communicating a great deal, and this is because most of them seem to have been social animals.

LIVING IN A HERD

Animals can be either solitary, living alone, or gregarious, living in groups. Today, many large carnivores, such as bears and leopards, are solitary, generally hunting alone. Also, by their modes of life, moles, skunks, and rhinos are often solitary. Most others live in herds.

We have seen the small groups of *Psittacosaurus* babies (see pages 182–3), evidence they were gregarious. Presumably they knew to be silent when a predator was near, and perhaps they gave one another warning squeaks or tail waves when they spotted danger.

Many dinosaur species are found in great numbers in single locations, and it's tempting to say this proves they lived in a herd. However, paleontologists are trained to question everything: Can we be sure the mass of *Triceratops* skeletons in a single locality couldn't be a result of a storm, when carcasses were washed into a pile by a flooding river?

Stronger evidence for herd living comes from track sites, where multiple fossil trackways may run side by side, or crisscross in ways that suggest the animals were all running around at the same time (see page 84). But, even in cases of fossil tracks we have to be careful—are we sure each track was made on the same day by a different animal, or could they represent, at an extreme, a single animal marching back and forth over a favorite patch of feeding ground?

Why live in a herd? There are many benefits: young animals can be protected in the middle of the herd; most animals can graze safely if there are sentries posted to warn the herd of danger; a group can fight back against a predator more easily than one; when they are on the move the herd can find food more readily. There are downsides: for example, a herd of herbivores would quickly strip all the vegetation from one spot and have to move on. We have already seen that many dinosaurs might have migrated thousands of miles each year as the seasons changed and food supplies became richer in one place or another.

HEADS AND FEATHERS AS VISUAL SIGNALS

Within the herd, all the dinosaurs would have had their individual features, but also there might have been differences between males and females (see page 167), and between youngsters and adults. These differences might have been clear to see in the feathers or head crests of different species, just as in some birds today.

Best known are the remarkable head crests of hadrosaurs, which ranged from no crest at all to an upended dinnerplate, a spike, a curved spine, and all sorts of other headgear. These were likely highly colored in life, and they gave other hadrosaurs immediate information. In addition, the crests were used as trumpets (see page 194).

BABY TO ADULT
In the baby *Parasaurolophus cyrtocristatus* (front), the head crest is short and the baby could probably only make high-pitched squeaking sounds when it puffed through its nose. But the adult (back) has a longer crest, and the air circulated through the whole length when it snorted, producing a deep-throated trumpeting noise.

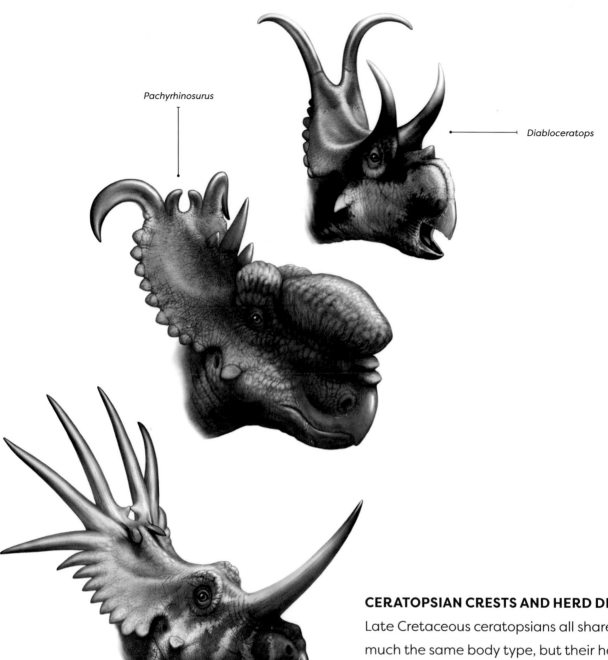

Pachyrhinosurus

Diabloceratops

Styracosaurus

CERATOPSIAN CRESTS AND HERD DEFENSE

Late Cretaceous ceratopsians all shared pretty much the same body type, but their headgear differed. In some localities, we find skeletons of three or four species together, so they may have fed in mixed herds, as some deer and antelope do today on the African plains.

The multiple horns on ceratopsian heads maybe had two functions: communication and defense. As each species had a different array of crest and horns, these are diagnostic and help us to identify the species. But they probably also helped the ceratopsians themselves recognize who was a member

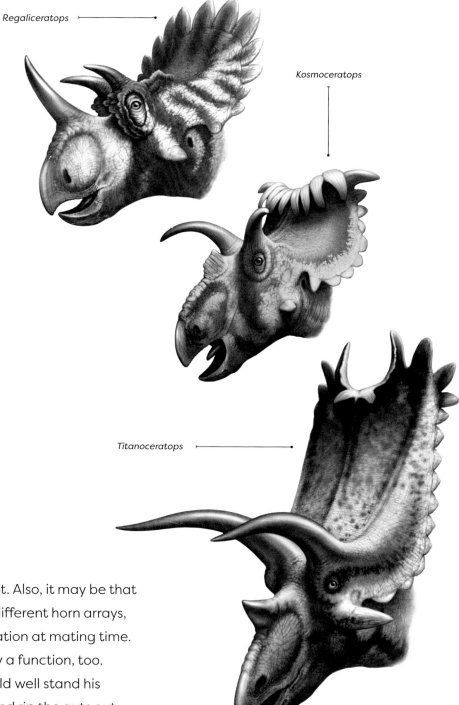

Regaliceratops

Kosmoceratops

Titanoceratops

FANCY HEAD GEAR
These Late Cretaceous ceratopsians from North America all have very similar skeletons, and many of them even lived side by side, but they show an amazing variety in their nose horns, forehead horns, and bony neck frills at the back.

of the same species or not. Also, it may be that males and females had different horn arrays, which helped communication at mating time.

Defense is obviously a function, too. An adult ceratopsian could well stand his ground when attacked and rip the guts out of a predator that came too close to its sharp horns. Also, based on the behavior of modern buffalo and musk oxen, it's been suggested that the feisty adults might have formed a ring or a row facing the predator attacked and kept circling, horns forward, as he prowled back and forth looking for a weak point to break through.

SOCIAL BEHAVIOR

FORENSICS:
SOUNDS OF THE HADROSAURS

Paleontologists are confident they know the noises hadrosaurs made. This is because their head crests were built from their nasal bones, and so the breathing tubes passed up and around through the crests.

For example, in CT scans of the skulls of *Corythosaurus*, David Evans, Larry Witmer, and colleagues of Ohio University showed that juveniles had shorter nasal passages than adults, and that the young animals made higher-pitched squeaks than the adults. Not only that, but membranes around the breathing tubes in the nostrils could swell up and probably helped to tune the sounds of their voices, perhaps adding extra vibration or making the sounds travel further.

In fact, this function in communicating by sound had been suggested a hundred years ago and studied in more detail by paleontologist David Weishampel in the 1980s. He looked at the hadrosaur *Parasaurolophus*, the one with a long tubular crest extending back from the top of its head. It's been shown that the crest is small in juveniles, and sprouts longer and longer as the animal ages, and might even differ in shape between males and females.

Weishampel traced out the convoluted course of the air passages through the *Parasaurolophus* crest and found the shape resembled that of a wind instrument like a trumpet, and in particular an ancient German instrument called a krummhorn. The krummhorn makes a low-pitched, rather rude-sounding parping noise that can carry for long distances, and

INSIDE THE DINOSAUR HEAD
These images, made from X-ray scans of hadrosaur skulls, show the brain (purple) and the nasal passage (green). This proves that the hadrosaur crests housed complex nasal passages, and experiments show that their twists and turns amplified sound and acted like trumpets, making honking and parping noises.

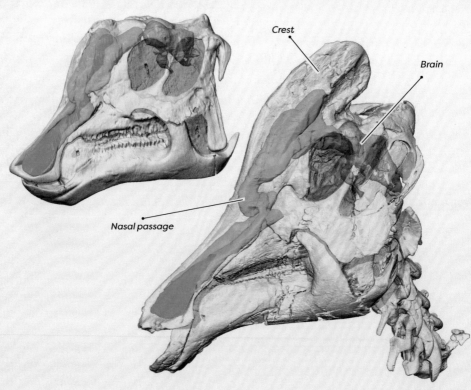

the investigators proposed this was the same for *Parasaurolophus*. In a herd, the males and females parped and boomed, perhaps at different levels, perhaps uttering a series of short and long toots to convey particular messages. The babies tweeted and tooted at a higher pitch.

Weishampel tested this idea by making a life-size model with cardboard and plastic. Computer scientists at Sandia National Laboratories went one step better and took the 3D nasal tube scans and ran them through a computer program designed to test musical instrument design. The computer then simulated different amounts of air being puffed through the tubes, as well as allowing for fleshy, moist skin. They confirmed the tooting and parping noises Weishampel had modeled, and you can hear their dinosaur sound reconstructions online at YouTube.

HADROSAUR HEAD CRESTS
The hadrosaurs had the most amazing array of head crests, some like long tubes (*Parasaurolophus walkeri*, top), others like hooked balloons (*Olorotitan arharensis*, middle), and others like a croissant on a stick (*Lambeosaurus lambei*, bottom).

SOCIAL BEHAVIOR

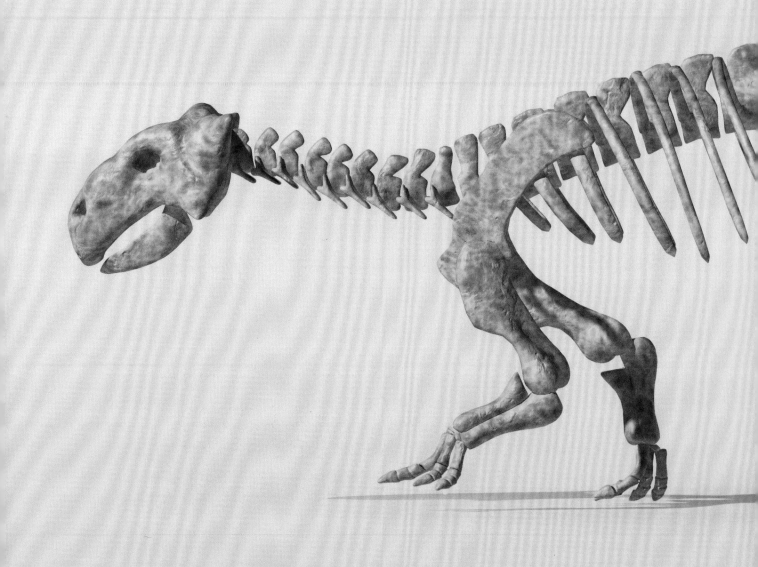

7
DINOSAURS AND HUMANS

EXTINCTION

Dinosaurs were killed by a huge meteorite impact 66 million years ago, but birds, mammals, and other reptiles survived—and took over the world.

For many people, the extinction is the main thing they know about dinosaurs. As this book has shown, however, there were so many interesting aspects of dinosaurs' lives when they dominated Earth. Their extinction opened the way for the rise of mammals, and humans are mammals. So, it's clear that we owe our present success on Earth to the demise of the dinosaurs 66 million years ago.

When we think about the extinction of dinosaurs, we should consider what killed them, and why they died out apparently so rapidly. The dinosaurs had been on Earth since the Early Triassic, over 180 million years earlier, and they had been dominant animals in their ecosystems for 160 million years. So, did they die out suddenly, or was something else going on? Was it a meteorite or great volcanic eruptions? How did Earth recover, and which survivors took over?

It's worth remembering that it wasn't just the dinosaurs that died out at the end of the Cretaceous. The flying pterosaurs (see page 24) and many marine reptiles including plesiosaurs and mosasaurs (see page 17) also died out. Many groups of mammals, birds, and lizards were also hit hard, although others survived. There were other victims among insects and plants on land, and among mollusks, corals, and fishes in the oceans. The abundant ammonites and belemnites disappeared.

IMPACT

One of the most amazing scientific discoveries of the last fifty years was that Earth was hit by a huge meteorite, or asteroid, measuring 7 miles (10 km) across, and weighing over a thousand billion tons (= ten to the power 12 tons). The evidence comes from the 150-mile (200-km) wide crater centered on Chicxulub, a small town in southeastern Mexico that dates to precisely the end of the Cretaceous, 66 million years ago.

There's also evidence of the catastrophe worldwide. As the asteroid struck, it punched down through Earth's crust, into the molten mantle, and vaporized. Then, there was a reaction to the huge amount of energy of the impact, and a back-blast of billions of tons of rock and dust shot high into the atmosphere. The back-blast formed a giant crater, and the rock began to fall back to Earth due to gravity.

Smaller debris did not fall back immediately but stayed high in the atmosphere for a few weeks. Two other things happened while great rocks the size of houses were falling back to earth. First, the asteroid had hit in the ocean and set off a huge tsunami wave that hurtled in all directions for hundreds of miles, creating storm beds when it hit the shores of Mexico and the United States. Second, the power of the impact melted bedrock, and millions of tons of glass beads flew out, small blobs of melt glass. As they spun through the air, they cooled, and eventually rained down over a vast area.

The dust in the high atmosphere, meanwhile, began to fall to earth in rainstorms, and this carried pulverized remains of the asteroid itself, which was rich in the rare metal iridium. Iridium is chemically akin to platinum and gold, but is even rarer on Earth, and arrives generally from space.

These four physical consequences of impact left their mark. Great boulders are seen

---

THE CHICXULUB BLAST

In the inner ring, the explosion was so intense that all organic matter (trees, animals) were vaporized. In a wider area the blast killed everything; farther out and all animals would have been deafened. Farthest of all, glass in buildings (if there had been any at the time) would have shattered.

 All organic material vaporized

 Lethal blast wave

 Eardrums damaged

 Glass shatters

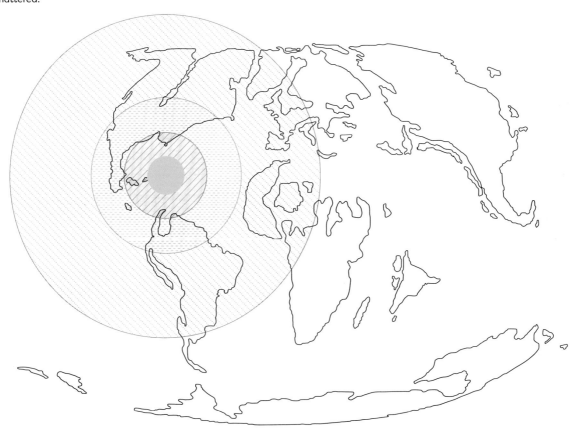

DINOSAURS AND HUMANS

PREVIOUS PAGE
The last day of the dinosaurs. A terrified ornithopod *Thescelosaurus* runs by and a pterosaur *Quetzalcoatlus* flaps in the sky. Fiery molten balls of glass are falling all around, thrown up by the impact many hundreds of miles away, and burning through the skin of these animals.

RIGHT
Fossil of the head of a paddlefish from the Tanis site in North Dakota.

around the crater site in Mexico. The tsunami beds are seen as great masses of chaotic boulders piled up on the coasts of Mexico and the southern United States. The melt glass beads are seen widely over the Caribbean and the United States as far north as the Dakotas. The dust layer with iridium marks the end of the crisis beds and is seen worldwide.

The killing was on a global scale. Close to the impact site, everything living was blasted away—trees, dinosaurs—all blasted and burned to nothing. Then, in a zone hundreds of miles wide everything was killed, but probably not burned. Even 2,000 miles (3,200 km) away the glass beads fell, there were blast waves, and Earth's surface shook, as documented at Tanis, North Dakota. Most of the killing happened in the weeks and months that followed, because a dust cloud blacked out the sun and caused the surface of the earth to become dark and very cold.

The iridium spike has been detected worldwide, in sediments laid down on land and under sea. This proves the dust fell everywhere, forming a blanket on land and sinking to the seabed. A wonderful rock section at Beloc on the island of Haiti shows each of the consequences of the impact: it begins with glass spherules, then a thick tsunami chaotic bed, and then is capped with a thin iridium-rich clay layer at the top.

THE LAST DAY OF THE DINOSAURS

In a powerful TV documentary in 2022, *Dinosaurs: The Final Day*, natural historian David Attenborough meets Robert De Palma, a remarkable young researcher who has spent years working at a site called Tanis in North Dakota. The rocks there belong to the Hell Creek Formation, an extensive rock unit over much of Montana, North Dakota, South Dakota, and Wyoming. There he had found dinosaur, turtle, mammal, and fish fossils all apparently killed on the very last day of the Cretaceous. This is quite a claim, but the case seems credible. De Palma can even

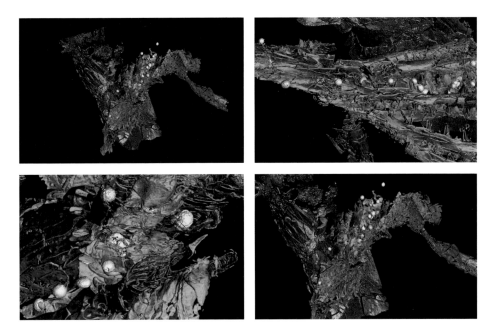

TERMINAL SPHERULES
Lit up as bright lights, the tiny glass beads, or spherules, are fallout from the great asteroid impact on the last day of the Cretaceous. Here, the false color from fluorescent lighting shows the bones of a fish (brown) and the cartilage.

tell that the asteroid hit in late May or early June.

Tanis was a coastal site at the time on the banks of the Western Interior Seaway, a great branch of the sea that split North America (see page 19). Evidence for the impact day includes marks in rock that suggest water ran both ways in these coastal rivers, probably set off by seismic waves that were generated by the Chicxulub impact. Seismic waves are associated with earthquakes, and they set the earth rattling and vibrating. Also, the fossil bed is full of glass spherules that match the Chicxulub site, and some are even said to contain remnants of the asteroid itself.

Did these shock waves and glass beads, as well as the later cooling and darkness, kill all the Tanis beasts? De Palma certainly thinks so, and he shows Attenborough all the evidence in the documentary. For example, the beads are found mixed with the bones of the dead animals, some even in the areas of the gills of large sturgeon fish. The suggestion is that these predatory fish, each about 3 ft (1 m) long were swimming lazily on a hot day, looking for food, when the burning hot glass beads began to rain from the sky. The beads sizzled as they fell into the water, and the fish swam away. But the beads kept falling, and the fish, gasping for air, swallowed great mouthfuls of burning glass beads that passed back into their gills and fried their flesh. The heat and pressure sucked the oxygen from the air and water surface; fish suffocated in rivers; the dinosaurs died on land.

In thin sections of sturgeon bones, Melanie During, a PhD student, identified growth lines suggesting seasonal growth, as often seen in dinosaurs (see page 188). As sturgeon are still alive today, During, De Palma, and colleagues were able to work out that the sturgeon they analyzed died at six years old, and near the beginning of the spring growth season, perhaps in May or June.

IMPACT OR ERUPTION?

There has been a long-running debate among geologists and paleontologists about the actual killer. Was it the asteroid impact or was it volcanic eruptions? There were huge eruptions in India, called the Deccan Traps, which cover 200,000 square miles (500,000 km^2) and represent billions of tons of lava spewed from volcanoes that began to erupt before the end of the Cretaceous and reached a peak of eruption near the time of extinction.

How do volcanoes kill? There have been many crises in Earth's history caused by such massive eruptions. It's not so much the lava that kills worldwide, but rather the gases pumped into the atmosphere. These are so-called greenhouse gases, such as carbon dioxide, methane, and water vapor, and their main effect is to raise temperatures. High temperatures kill life on land and in the oceans.

But we now know it was the Chicxulub impact, not the Deccan Traps, that was responsible for all or most of the killing. New evidence presented by paleontologist Celli Hull shows that the Deccan Traps reached peak eruption 200,000 years before the end of the Cretaceous, but most of the extinctions happened right at the end of the Cretaceous, just when the asteroid hit Earth.

VOLCANIC LANDSCAPE
The Deccan Traps in northern India as they are today, showing lava rocks and flowers near Koyna Lake, Western Ghats, during the monsoon. The huge eruptions just before the end of the Cretaceous led to a burst of global warming that led to some extinctions, but these were a precursor of the killer impact that happened 200,000 years later.

FORENSICS: A WHIMPER OR A BANG?

Did the end-Cretaceous extinctions happen in an instant, simply as a result of the Chicxulub impact? There is no question the impact was the killer, but two other things might have kicked off some extinctions earlier: the Deccan Traps volcanic eruptions, and the fact that the climate was changing and cooling.

Geologists can establish ancient air and sea surface temperatures by measuring oxygen isotopes in rocks and fossils. These oxygen isotope measurements can indicate something about the water and food animals were consuming at the time (see page 150), but averaging many measurements from the same age can give an indication of temperature. Through the Cretaceous, average global temperatures fell from hothouse levels of around 77–86°F to 59–68°F (25–30°C to 15–20°C), then continued falling to present-day levels of around 50–59°F (10–15°C).

Could it be that dinosaurs were on the decline anyway, perhaps because of the cooling temperatures and major changes to the landscape caused by the rise of flowering plants, which were not core parts of their diets (see page 152)? This is a controversial theory, but some recent studies suggest dinosaurs might have been past their best, if not already declining, for the last 40 million years of the Cretaceous.

These studies were performed using new mathematical methods called Bayesian statistics. In these methods, the analyst starts with a likely model, then varies all the uncertainties in the information available.

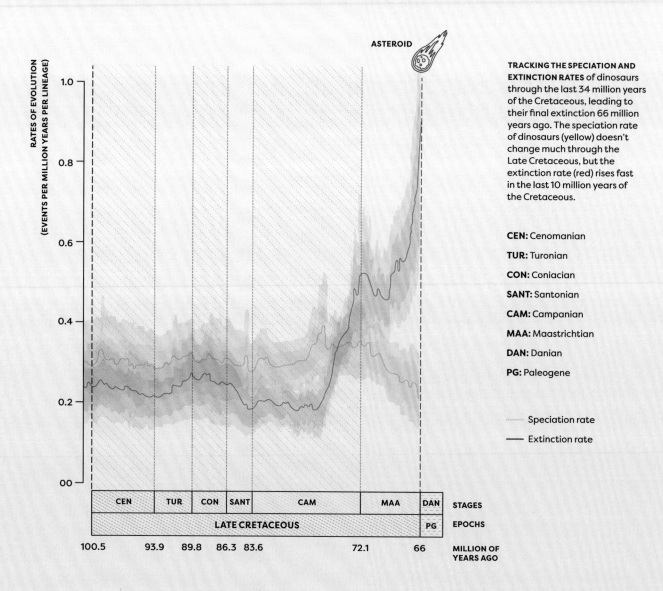

TRACKING THE SPECIATION AND EXTINCTION RATES of dinosaurs through the last 34 million years of the Cretaceous, leading to their final extinction 66 million years ago. The speciation rate of dinosaurs (yellow) doesn't change much through the Late Cretaceous, but the extinction rate (red) rises fast in the last 10 million years of the Cretaceous.

CEN: Cenomanian
TUR: Turonian
CON: Coniacian
SANT: Santonian
CAM: Campanian
MAA: Maastrichtian
DAN: Danian
PG: Paleogene

These uncertainties include the counts of fossil species, the dating of rocks, and gaps in the rock record data. The computer grinds away, sometimes for weeks, trying out every possible model that covers all the worries and uncertainties, and a final result emerges that shows the most likely result plus a cloud of all other possible results.

In studies of worldwide dinosaur records and data from North America alone, there was a definite decline in species numbers. The evidence shows that the rate of origin of new species dipped a little in the last 5 million years of the Cretaceous, but the species extinction rate increased dramatically. Not all groups were affected equally, and two groups, the hadrosaurs and ceratopsians, were surviving the climate changes well.

Also, you probably would not be aware of the decline if you hopped in a time machine and visited any of the famous Late Cretaceous dinosaur scenes, such as the Hell Creek Formation. These were full of species of dinosaurs, as well as all the other land animals we would expect from the era. But the total number of dinosaur species across the whole of North America, and even the whole of the world, was down. What we can't tell for sure is whether the number of individual animals was also going down.

TRACKING THE NET DIVERSIFICATION RATE (that is, speciation minus extinction) of dinosaurs through the last 34 million years of the Cretaceous, leading to their final extinction 66 million years ago. The sum of the speciation and extinction rates (see graph, left) adds up to a steady diversity of dinosaur species through the Late Cretaceous, and then a long-term, steep fall in the millions of years before the end of the Cretaceous and the impact of the asteroid.

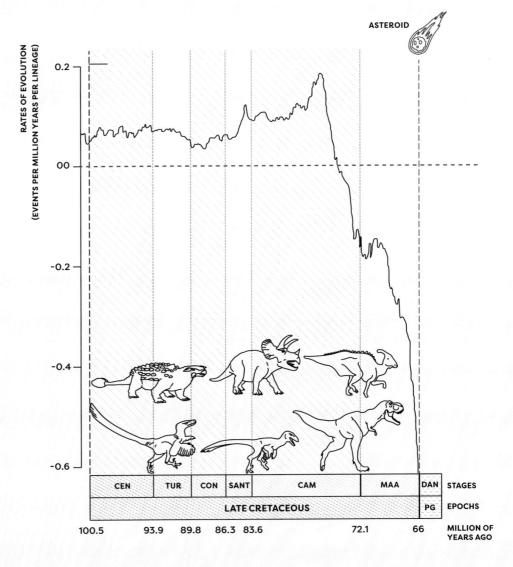

DINOSAURS AND HUMANS

SURVIVORS AND THE MODERN WORLD

From the point of view of humans, the extinction of the dinosaurs 66 million years ago finally opened the way for mammals to take over.

Humans belong to the order Primates, which also includes lemurs, monkeys, and apes. The origin of primates can be tracked back to the years just after the end of the Cretaceous, the time from 66–23 million years ago called the Paleogene. We often forget that mammals arose at about the same time as dinosaurs, in the Late Triassic, and the two groups lived side by side throughout the Jurassic and Cretaceous. These early mammals diversified into many groups with a variety of habits and diets, but none of them was larger than a cat.

New work shows that the evolutionary effect of the sudden disappearance of the dinosaurs was to release the mammals from their constraints. In a sense, dinosaurs were everywhere, doing all the "big animal" things, and mammals had to live on the margins, staying small, feeding at night. With the dinosaurs gone, there were rich opportunities for mammals, and for other groups such as birds and crocodilians.

In the Paleogene, there were all sorts of amazing new animals, as these three groups evolved fast to fill all the opportunities.

SURVIVORS
Mammals and birds lived side by side with the dinosaurs in the Jurassic and Cretaceous, but some of them also survived through the extinction. Here are some amazing fossil birds from the last 66 million years, the huge seabird *Dasornis*, the giant penguin *Palaeeudyptes*, and the plant-eating flightless relative of modern chickens and ducks called *Gastornis*.

In South America, the giant bird *Phorusrhacos*, nearly 9 ft (2.7 m) tall, fed on sheep-sized mammals. Giant crocodilians became more adapted to life on land and were top predators in other regions. It took a while for the mammals to establish their dominance on land.

The typical scene within a few million years after the big extinction looked so different. The angiosperms, which arose in the Cretaceous (see page 152), had taken over, and with no dinosaurs to crop the trees and bushes, they flourished. The first tropical rain forests soon developed in the warmer areas. These were full of early relatives of familiar modern groups such as horses, cattle, shrews, rats, cats, dogs, and monkeys. There were also some weird experiments, large mammals with horns, that did well but did not survive to the present day.

The first primates looked like rats, small with long tails, and scuttling in the trees hunting insect prey. Some such as *Plesiadapis* were larger, with long fluffy tails, and strong hands and feet for grasping branches. These were herbivores that fed on leaves. During the Paleogene, primates split into ancestors of modern lemurs and tarsiers on the one hand, and monkeys and apes on the other. Some monkeys got across to South America, where they evolved their own features, including a grasping tail that they use as a fifth limb.

In Africa and Europe, another group of monkeys lost their tails and had stronger arms for swinging through the trees. These were the apes, which gave rise to modern gorillas, chimps, and humans, about 5 million years ago.

Dasornis

Palaeeudyptes

Gastornis

OPPORTUNITIES

Humans are connected to dinosaurs through the evolution of mammals and their prominence after dinosaur extinction but also in public enthusiasm, ideas, and jobs.

How do we understand the world around us? For thousands of years, thoughtful people have looked, discussed, and tried to understand. Why does the sun rise in the same place each day? How can we grow plants to eat? Why do people get sick? For the last two hundred years, these questions have been identified as parts of science, and the pursuit of scientific knowledge has become professional.

The people who study new ideas in medicine, agriculture, and how the universe works have to work within strict rules. The first rule is honesty: you can't make things up, and you have to present evidence for any new idea. Linked to this is repeatability: if you make a claim ("I've found a new dinosaur"; "I've invented a cure for cancer"), you not only have to report it honestly and provide evidence, but also the evidence has to be available so others can check it. So, there are whole systems in professional science that are agreed worldwide to try to control the boundaries between real science and nonsense.

PROFESSIONAL STANDARDS

Research results have to be published in peer-reviewed journals; it's like a court of law—you have to convince a small committee of experts that your claims and evidence are honest, that the work can be checked and repeated, and that you aren't overstating the case. "Peer review" means just that, the process of review or criticism by your peers, meaning other professional people who are your equals.

These rules are important because they mean that you can trust some sources of data, and you learn to be careful about others. All the results and conclusions mentioned in this book are based on careful research that has been through the peer-review process, and most news resources such as magazines and newspapers, and science websites such as Wikipedia try to respect these rules.

You might read on a web page or in a blog about a new dinosaur that has five legs, or a miracle product that will cure every kind of illness. But, check closely: is a professional article referenced anywhere? Click the links at the end and see what the evidence is.

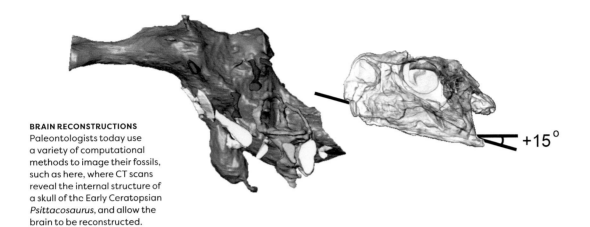

BRAIN RECONSTRUCTIONS
Paleontologists today use a variety of computational methods to image their fossils, such as here, where CT scans reveal the internal structure of a skull of the Early Ceratopsian *Psittacosaurus*, and allow the brain to be reconstructed.

The professional trust in science means that scientists require thorough training, and it is possible to get a paid job as a professional paleontologist in many universities or museums. We'll look at how you get there.

JOBS IN THE LAB

We've already looked at the practical aspects of digging up a dinosaur (see pages 30–1). And digging it up is only the start. The skeleton may be large (of course) and divided into multiple blocks of rock with bones, each weighing up to a ton. These require careful transportation and handling. Then there may be many months of work to extract bones from rocks and prepare them for study.

Not all dinosaur skeletons are put on show. All the great museums of the world already have such displays, and new fossils might not be as good or might be another example of a dinosaur that is already on show. Practical lab work is done by paleontology preparators, skilled technical staff who have a training in paleontology, anatomy, and a variety of equipment. They have to know which drills to use to remove rock, how to protect bones from damage using chemicals, and how to make casts and molds.

If bones are for display, a whole bunch of additional technical skills are needed, to identify the bones correctly and work out how to build an armature, the metal or plastic frame on which the bones are organized to form a skeleton. If it's the original bones, they may weigh tons, and skillful engineering is needed to make sure the whole construction does not collapse.

Then there are paleoartists who may make paintings or 3D models to show the animals in life appearance. These technical experts require artistic ability and a training in anatomy, how animals move, and how to construct models with different materials. They also work with paleontologists to make sure images and models are accurate (see page 51).

There are some tools paleontologists use a great deal. Most important in the past thirty years has been the use of CT scanning (see pages 108–09). Lab technicians are trained to operate the scanner to get the best results. The scanners are of medical grade, and generally use X-rays to look inside a block of rock and even inside bones. Detail can be good enough to see all of the bone construction, including LAGs (see page 188). Scans are also the basis for 3D digital models of skulls and skeletons, which are the first step not only in making reconstructions, but also in calculating physical aspects, such as bite forces of dinosaur jaws (see page 141) or modes of locomotion (see pages 78–9).

NEW RESEARCH

It's often assumed that the main job for paleontologists is to dig up more and more fossils. This is important, and we have to identify which species lived at different times and where they lived. However, it's now possible to learn so much about *how* the dinosaurs lived too, in other words, their paleobiology. A few years ago, some of these kinds of studies would have been just

PRECISION DRILLING
Dinosaur bones are usually encased in rock and lab technicians spend a lot of time using drills to remove the rock carefully and expose the detail of the bones.

guesswork or speculation. But now we can even reconstruct dinosaur feather colors and the sounds some of them made.

As a quick reminder of the kinds of studies in this book, first we have methods in physics, and particularly in biomechanics. In studying how dinosaurs moved about and how their jaws worked, we can now use standard mechanical and engineering methods to test what ancient animals could do. Some of the discoveries are on an epic scale because dinosaurs and other ancient animals were doing things that no living animal does. For example, *T. rex* had a bite force about ten times as powerful as the great white shark—we know this and we can test it. Some giant pterosaurs were ten times as big as the largest birds today, but we know how they flew.

We can use chemistry to identify ancient climates and diets. For example, oxygen and carbon isotopes provide a lot of information about atmospheres and temperatures (see page 206). Oxygen and nitrogen isotopes can help identify what ancient animals were eating (see page 150). Chemistry is also useful in studying the decay and preservation of fossils, and can be used to identify ancient pigments that gave colors in skin and feathers (see pages 172–3).

We can use mathematics to work out patterns and processes of evolution. For example, in studies of rates of evolution around mass extinctions (see pages 34 and 198–207), paleontologists need detailed data on all the known dinosaur fossils through the Cretaceous, their dates, and names. They can then use this information to work the rates of formation of new species, extinction, and preservation quality. In looking at the evolution of function, paleontologists can combine engineering analyses of ancient jaws (see page 141) with the evolutionary tree to work out when and how particular adaptations arose.

PALEONTOLOGIST CAREERS

There's so much we don't know. In the future, paleontologists who are kids now will become professionals and may discover the first dinosaurs from the Early or Middle Triassic. They will understand better how the giant dinosaurs operated their body systems. They will learn exactly how bird flight originated, how dinosaur ecosystems worked, and exactly how the end-Cretaceous mass extinction happened. Many of these questions have profound meaning for how we understand big themes such as origins, biodiversity, climate change, extinction, and conservation.

These big science questions and all the amazing new methods that have been developed in the last twenty years mean there is a need for paleontological science research. Also, public interest is significant, and this means there's demand for movies, books, documentaries, and museum displays. In turn, these demand new science, and new science needs new scientists.

Paleontology is open to everyone. You don't have to be a hardy field worker; most of the important research is done sitting at a computer. Students have to study all the core sciences (biology, physics, chemistry) as well as mathematics and geology. Many of the jobs require at least a bachelor's degree in Biology, Geology, or Paleontology. Many jobs might also require a master's degree in Paleontology, Museum Studies, or Fossil Preparation. Research jobs in universities and major museums also require a doctorate. There are many websites that give detailed advice on the studies you need.

If you love dinosaurs, visit your local museum, join a dig, and speak to any professional paleontologists you can find. These are exciting times for all areas of science, and most nations want to encourage as many young people as possible to consider careers of this kind. The more you learn about dinosaurs and how they lived, the more you realize we don't know!

RIGHT
Paleontology is a great career for people who enjoy practical, hands-on work and interacting with kids and the public. In the lab (top), the technician is working on a delicate fossil using a drill to remove unwanted rock debris. The fossil is in the box so dust can be extracted through the vacuum tube system. Visitors love to be at the heart of the action (below), and many museums have their dino lab on show as an exhibit, so the paleontologists can explain what they are doing.

NEXT PAGE
A scene in the Middle Jurassic of England, showing a theropod dinosaur, *Cruxicheiros*, sniffing the air (left) and heading for a corpse of the marine crocodile *Steneosaurus* that has washed up on the beach. Theropods were hunters, but like modern carnivores they would also take advantage of a free lunch if they had the chance.

BIBLIOGRAPHY

CHAPTER 1: DINOSAURS IN PERSPECTIVE

Barker, C. 2020. *The Ultimate Dinosaur Encyclopedia.* Wellbeck, London.

Benton, M.J. 2019. *The Dinosaurs Rediscovered. How a Scientific Revolution is Rewriting History.* Thames & Hudson, New York, London.

Benton, M.J. 2019. *Cowen's History of Life.* Wiley, New York.

Brusatte, S. 2018. *The Rise and Fall of the Dinosaurs.* Macmillan, London.

Fastovksy, D.E. & Weishampel, D.B. 2021. *Dinosaurs: A Concise Natural History.* Cambridge University Press, Cambridge.

Lomax, D.R. 2021. *Locked in Time. Animal Behavior Unearthed in 50 Extraordinary Fossils.* Columbia University Press, New York.

Naish, D. 2023. *Ancient Sea Reptiles: Plesiosaurs, Ichthyosaurs, Mosasaurs and More.* Natural History Museum, London.

Witton, M.P. 2013. *Pterosaurs: Natural History, Evolution, Anatomy.* Princeton University Press, Princeton.

CHAPTER 2: PHYSIOLOGY

Benton, M.J. 2021. *Dinosaurs. New Visions of a Lost World.* Thames & Hudson, New York, London.

Benton, M.J., Dhouailly, D., Jiang, B., & McNamara, M. 2019. The early origin of feathers. *Trends in Ecology & Evolution* 34, 856–869.

Ksepka, D.T. 2020. Feathered dinosaurs. *Current Biology* 30, R1347–R1353.

White, S. & Naish, D. 2022. *Mesozoic Art: Dinosaurs and Other Ancient Animals in Art.* Bloomsbury Wildlife, London.

Woodruff, C. & Wolff, E. 2022. Sauro-throat. *The Linnean* 38, 9–13.

Xing, L., McKellar, R.C., Xu, X., Li, G., Bai, M., Persons, W.S., Miyashita, T., Benton, M.J., Zhang, J., Wolfe, A.P., & Yi, Q. 2016. A feathered dinosaur tail with primitive plumage trapped in mid-Cretaceous amber. *Current Biology* 26, 3352–3360.

CHAPTER 3: LOCOMOTION

Alexander, R.M. 1976. Estimates of speeds of dinosaurs. *Nature* 261, 129–130.

Biewener, A.A. & Patek, S.N. 2018. *Animal Locomotion.* Oxford University Press, Oxford, New York.

Brusatte, S.L., O'Connor, J.K., & Jarvis, E.D. 2015. The origin and diversification of birds. *Current Biology* 25, R888–R898.

Chiappe, L.M. 2007. *Glorified Dinosaurs: The Origin and Early Evolution of Birds.* Wiley, New York.

Gatesy, S.M., Middleton, K.M., Jr, F.A.J., & Shubin, N.H. 1999. Three-dimensional preservation of foot movements in Triassic theropod dinosaurs. *Nature* 399, 141–144.

Klein, N., Remes, K., Gee, C.T., & Sander, P.M., eds. 2011. *Biology of the Sauropod Dinosaurs: Understanding the Life of Giants.* Indiana University Press, Bloomington.

Richter, A. & Falkingham, P.L. 2016. *Dinosaur Tracks: The Next Steps.* Indiana University Press, Bloomington.

Sander, P.M., Christian, A., Clauss, M., Fechner, R., Gee, C.T., Griebeler, E.M., Gunga, H.C., Hummel, J., Mallison, H., Perry, S.F., & Preuschoft, H., 2011. Biology of the sauropod dinosaurs: the evolution of gigantism. *Biological Reviews* 86, 117–155.

Xu, X., Zhou, Z., Dudley, R., Mackem, S., Chuong, C.M., Erickson, G.M., & Varricchio, D.J. 2014. An integrative approach to understanding bird origins. *Science* 346, 1253293.

CHAPTER 4: SENSES AND INTELLIGENCE

Balanoff, A.M., Bever, G.S., Rowe, T.B., & Norell, M.A. 2013. Evolutionary origins of the avian brain. *Nature* 501, 93–96.

Ballell, A., King, J.L., Neenan, J.M., Rayfield, E.J., & Benton, M.J. 2021. The braincase, brain, and palaeobiology of the basal sauropodomorph dinosaur *Thecodontosaurus antiquus*. *Zoological Journal of the Linnean Society* 193, 541–562.

Brusatte, S.L., Norell, M.A., Carr, T.D., Erickson, G.M., Hutchinson, J.R., Balanoff, A.M., Bever, G.S., Choiniere, J.N., Makovicky, P.J., & Xu, X. 2010. Tyrannosaur paleobiology: new research on ancient exemplar organisms. *Science* 329, 1481–1485.

Buchholtz, E. 2012. *Dinosaur Paleoneurology.* Indiana University Press, Bloomington.

Choiniere, J.N., Neenan, J.M., Schmitz, L., Ford, D.P., Chapelle, K.E., Balanoff, A.M., Sipla, J.S., Georgi, J.A., Walsh, S.A., Norell, M.A., & Xu, X. 2021. Evolution of vision and hearing modalities in theropod dinosaurs. *Science* 372, 610–613.

Ksepka, D.T. 2021. Bird brain evolution. *American Scientist* 109, 352–360.

Parrish, J.M., Molnar, R.E., Currie, P.J., & Koppelhus, E.B. eds. 2013. *Tyrannosaurid Paleobiology.* Indiana University Press, Bloomington.

CHAPTER 5: FEEDING

Barrett, P.M. 2014. Paleobiology of herbivorous dinosaurs. *Annual Review of Earth and Planetary Sciences* 42, 207–230.

Barrett, P.M. & Rayfield, E.J. 2006. Ecological and evolutionary implications of dinosaur feeding behaviour. *Trends in Ecology & Evolution* 21, 217–224.

Button, D.J., Rayfield, E.J., & Barrett, P.M. 2014. Cranial biomechanics underpins high sauropod diversity in resource-poor environments. *Proceedings of the Royal Society B* 281, 20142114.

Chin, K., Tokaryk, T.T., Erickson, G.M., & Calk, L.C. 1998. A king-sized theropod coprolite. *Nature* 393, 680–682.

Rayfield, E.J. 2007. Finite element analysis and understanding the biomechanics and evolution of living and fossil organisms. *Annual Review of Earth & Planetary Sciences* 35, 541–576.

Rayfield, E.J., Norman, D.B., Horner, C.C., Horner, J.R., Smith, P.M., Thomason, J.J., & Upchurch, P. 2001. Cranial design and function in a large theropod dinosaur. *Nature* 409, 1033–1037.

Schaeffer, J., Benton, M.J., Rayfield, E.J., & Stubbs, T.L. 2020. Morphological disparity in theropod jaws: comparing discrete characters and geometric morphometrics. *Palaeontology* 63, 283–299.

Wang, S., Stiegler, J., Amiot, R., Wang, X., Du, G.H., Clark, J.M., & Xu, X. 2017. Extreme ontogenetic changes in a ceratosaurian theropod. *Current Biology* 27, 144–148.

Wolff, E.D., Salisbury, S.W., Horner, J.R., & Varricchio, D.J. 2009. Common avian infection plagued the tyrant dinosaurs. *PLoS One* 4(9), e7288.

CHAPTER 6: SOCIAL BEHAVIOR

Carpenter, K., Hirsch, K.F., & Horner, J.R. eds. 1996. *Dinosaur Eggs and Babies*. Cambridge University Press, Cambridge, New York.

Chiappe, L.M., Marugán-Lobón, J., Ji, S.A., & Zhou, Z. 2008. Life history of a basal bird: morphometrics of the Early Cretaceous *Confuciusornis*. *Biology Letters* 4, 719–723.

Erickson, G.M., Makovicky, P.J., Currie, P.J., Norell, M.A., Yerby, S.A., & Brochu, C.A. 2004. Gigantism and comparative life-history parameters of tyrannosaurid dinosaurs. *Nature* 430, 772–775.

Horner, J.R. 1984. The nesting behavior of dinosaurs. *Scientific American* 250(4), 130–137.

Lockley, M.G., McCrea, R.T., Buckley, L.G., Deock Lim, J., Matthews, N.A., Breithaupt, B.H., Houck, K.J., Gierliński, G.D., Surmik, D., Soo Kim, K., & Xing, L. 2016. Theropod courtship: large scale physical evidence of display arenas and avian-like scrape ceremony behavior by Cretaceous dinosaurs. *Scientific Reports* 6(1), 18952.

Mallon, J.C. 2017. Recognizing sexual dimorphism in the fossil record: lessons from nonavian dinosaurs. *Paleobiology* 43, 495–507.

Norell, M.A., Clark, J.M., Chiappe, L.M., & Dashzeveg, D. 1995. A nesting dinosaur. *Nature* 378, 774–776.

Saitta, E.T. 2015. Evidence for sexual dimorphism in the plated dinosaur *Stegosaurus mjosi* (Ornithischia, Stegosauria) from the Morrison Formation (Upper Jurassic) of western USA. *PloS One* 10(4), e0123503.

Vinther, J. 2020. Reconstructing vertebrate paleocolor. *Annual Review of Earth and Planetary Sciences* 48, 345–375.

Zhang, F., Kearns, S.L., Orr, P.J., Benton, M.J., Zhou, Z., Johnson, D., Xu, X., & Wang, X. 2010. Fossilized melanosomes and the color of Cretaceous dinosaurs and birds. *Nature* 463, 1075–1078.

Zhao, Q., Benton, M.J., Xu, X., & Sander, P.M. 2013. Juvenile-only clusters and behavior of the Early Cretaceous dinosaur *Psittacosaurus*. *Acta Paleontologica Polonica* 59, 827–833.

CHAPTER 7: DINOSAURS AND HUMANS

Black, R. 2022. *The Last Days of the Dinosaurs*. History Press, New York.

Brusatte, S.L., Butler, R.J., Barrett, P.M., Carrano, M.T., Evans, D.C., Lloyd, G.T., Mannion, P.D., Norell, M.A., Peppe, D.J., Upchurch, P., & Williamson, T.E. 2015. The extinction of the dinosaurs. *Biological Reviews* 90, 628–642.

Condamine, F.L., Guinot, G., Benton, M.J., & Currie, P.J. 2021. Dinosaur biodiversity declined well before the asteroid impact, influenced by ecological and environmental pressures. *Nature Communications* 12(1), 3833.

During, M.A., Smit, J., Voeten, D.F., Berruyer, C., Tafforeau, P., Sanchez, S., Stein, K.H., Verdegaal-Warmerdam, S.J., & van der Lubbe, J.H. 2022. The Mesozoic terminated in boreal spring. *Nature* 603, 91–94.

Gregory, J. 2019. *Paleontologist*. AV, New York.

Vangelova, L. 2019. Paleontologist. *The Science Teacher* 86, 64–65.

INDEX

A

abelisaurids 136
Acheroraptor 71, 134, 135
Acrocanthosaurus atokensis 115
Africa 16, 18, 19
age, estimating 186, 188
Alamosaurus 85
Alberta 19, 144
Albertosaurus 189
Alexander formula 89
Alioramus altai 115
alligators 52, 53, 54, 118, 181
allosaurids 38, 136
Allosaurus 94, 135, 136, 141
alvarezsaurids 125, 134, 161
amber fossils 61-3
Anchiornis 98, 101, 170, 171, 172
angiosperms 62, 152, 209
ankylosaurs 37, 39, 83, 84, 114, 121, 138, 161
Ankylosaurus 70, 71
Anseriformes 115
Antarctica 16, 18, 19
Apatosaurus 67, 92, 186, 188
Archaeopteryx 40, 42, 98, 99, 102-3, 112, 113, 114, 115, 128, 130, 179
Archimedes Principle 93-4
Arctic Ocean 16
Argentina 32, 153, 179
Argentinosaurus 95
Asia 16, 19, 62
asteroids 198-9, 202, 203, 204, 206, 207
Attenborough, David 202, 203
Auca Mahuevo, Argentina 179
Australia 16, 18, 19, 82, 83
Australovenator 82
Aves 115
Avialae 115

B

babies 181-2, 184-7, 188, 189
Bajadasaurus 11
Bakker, Robert T. 86
balance 66, 110, 126-31
Balanoff, Amy 114
Barosaurus 81, 136
Baryonyx 160, 161
Bayesian statistics 206-7
Bellusaurus 151
biomechanics 88-91, 134, 213

bipedal posture 47, 66, 79, 81, 83, 92, 126-7
Bird, Roland T. 86
birds
    body temperatures 53, 54
    brain 107, 110, 111, 112, 113, 114, 115
    breathing 64-5
    digestion 144-5
    extinction 198
    feathers *see* feathers
    lekking behavior 175
    locomotion 78-9
    origins 25, 35, 37, 38
    posture 76, 77
    senses 125, 130, 131
    as theropods 40-3
bite force 141, 212, 213
bite marks 143
body mass
    and brain size 116, 118
    endotherms 53, 54, 56
    estimating 94-5
    and flight 98, 99
    and growth rates 186, 188-9
body temperatures 52-7
bones 54, 55-7, 65, 188, 211
Borealopelta 144, 146
Brachiosaurus 92-3, 94, 138, 145, 154, 155
brains 106-19, 211
Brasier, Martin 106-7
breathing 64-7
Brontosaurus 11, 54, 55, 66, 92, 93, 94, 138, 179
Bruhathkayosaurus 94
Bullar, Claire 127
Button, David 154

C

Camarasaurus 92, 93, 138, 154, 155
Canada 19, 34, 144
Caprimulgiformes 115
Carboniferous 12, 178
carcharodontosaurids 136
Carcharodontosaurus 43
careers 210-14
Caribbean 16, 19, 62, 84, 202
carnivores
    behavior 158-61, 190
    evidence of diet 134, 144, 145, 150, 151

food pyramid 69
jaw measurements 158
size of 42, 135-7
teeth 138, 140
Caudipteryx 51, 60, 171
Cenozoic 14
Cerapoda 39
ceratopsians 37, 39, 81, 83, 114, 161, 192, 193, 207
Ceratosauria 38
chewing 138, 139-40
Chicxulub, Mexico 198-9, 202, 203, 206
Chin, Karen 146, 147
China 35, 40, 51, 58
Chinsamy-Turan, Anusuya 169, 170
Choiniere, Jonah 125
Citipati 115, 178-9
cloaca 174
coelophysids 135
Coelophysis 15, 23, 78, 79, 141
Coelurosauria 115
Colbert, Ned 93, 94
cold-blooded animals 22, 34, 52-7, 186
Colorado 19, 86, 174, 175
Columbiformes 115
communication 190-5
Conchoraptor gracilis 115
Confuciusornis 168-9, 172
continental drift 16
Cope, Edward 42
coprolites 146-7, 149, 152, 174
Coraciiformes 115
Corythosaurus 194
courtship 166, 167, 174-5
creches 182-3
Cretaceous 12, 14, 16, 18, 19, 29, 34, 37, 38, 62, 70, 71, 82, 84, 152, 198, 203, 204, 206-7, 208
crocodiles 52-3, 54, 57, 65, 78, 84, 86, 96, 124, 131, 145, 158, 179, 181, 209
Cruxicheiros 214
Crystal Palace display 47, 49
CT scanning 108-9, 211-12

D

Dakotaraptor 70, 135
Darwin, Charles 166, 167
Dasornis 208, 209
Daspletosaurus 189

De Palma, Robert 143, 202, 203
Deccan Traps, India 204, 206
Deinonychosauria 38, 115
Deinonychus 40, 43, 101, 126, 161, 177
Denali Natural Park, Alaska 85
Diabloceratops 192
Dicraeosaurus 154, 155
diet
    carnivore 134, 135-7, 158-61
    changes in 151, 161
    and diseases 147-9
    evidence of 142-7, 150-1
    food budgets 68-71
    and growth stages 136-7
    herbivore 134-5, 152-5
    and jaws 141, 142, 158-9
    omnivore 134, 150, 151
    specialist 161
    and teeth 138-40, 142
dig sites 30-1
digital images 49, 51
digitigrade stance 36
Dinosauria 36
dinosaurs
    brain 110, 111, 113, 114-19
    breathing 64
    diet and size 134, 135
    diversification rates 207
    evolution of 36-9
    extinction 198-207, 208
    feathers 59, 60
    mating 166, 167, 174-5
    origins 32-5
    posture 77
    senses 120-31
    sexual selection 166-71
    speciation rates 206
Dinosaurs: The Final Day 202
Diplodocus 11, 69, 92, 94, 138, 154, 155
diversification rates 207
Dromaeosauridae 115
dromaeosaurs 135
duckbilled ornithischians 139
During, Melanie 203

E

Early Cretaceous 82
ears 125-31
Earth 12-15, 16-19
ectotherms 52-7
Edmontosaurus 25, 71, 142
eggs 176-9, 181, 184, 185

elephants 11, 42, 54, 69, 86, 88, 92, 96, 113, 140, 144, 184, 188
encephalization quotient (EQ) 112-13, 114, 116
endocasts 106, 108, 109, 115, 129
endotherms 52-7, 64-5, 69, 96, 188
endurance 34, 77
energy flows 68, 69
Eoraptor 32, 33, 36, 81
Erickson, Greg 188
Eudromia elegans 78, 79
Euoplocephalus 11
Euornithopoda 39
Europe 12, 16, 18, 19
Evans, David 194
extinction 33, 34, 198-207
extinction survivors 208-9
eyes 122, 123

F

Falconiformes 115
FEA 141, 154, 161
feathers
  amber fossils 61-3
  color of 51, 58, 160, 171, 172-3, 213
  origins 35, 40
  role 24, 25, 58-9
  types 59-60
field of vision 122
finite element analysis 141, 154, 161
fish 110, 111, 112, 140, 161, 198, 202, 203
flight 24, 40, 42, 43, 59, 98-103
food budgets 68-71, 134
  see also diet
footprints 82-91
fossils 28, 29, 30, 31, 61-3

G

gait 76-81, 90-1
Galliformes 115
Gansus zheni 144
Gastornis 208, 209
gastroliths 142, 144-5
Gatesy, Stephen 90
Gaviiformes 115
Genasauria 39
geological time 10, 12-15

Giganotosaurus 43, 121
Gigantoraptor 10
gigantothermy 97
Giraffatitan 92
gliding 98, 101, 102, 103
global temperatures 206
Gondwana 16
Gorgosaurus 189
Grand Canyon, Arizona 12
growth rates 136-7, 184-9, 203
Gruiformes 115
Guanlong 151

H

hadrosaurs 71, 85, 139-40, 152, 161, 194-5
Hawkins, Benjamin Waterhouse 47, 49
head crests 166, 191-3, 194, 195
hearing 110, 124-5
Hell Creek Formation, Montana 29, 70, 71, 134, 202
herbivores
  diet 69, 134-5, 144, 145, 150, 151, 152-5
  growth rates 136, 137
  teeth/jaws 138, 158
herds 84, 86, 190-3
Herrerasauridae 38
Herrerasaurus 32, 33, 36
Heterodontosauridae 39
Heyuannia 177
hind legs, evolution of 36-7
Hitchcock, Edward 83
Horner, Jack 181
Hull, Celli 204
hunting 159-61

I

Ichthyornis 130
ichthyosaurs 21, 22, 23, 33
Ichthyovenator 10
Iguanodon 46, 47, 49, 84, 86, 107
Incisivosaurus gauthieri 115
India 16, 18, 19, 204, 206
intelligence 106, 110, 111-19
Isaberrysaura 153
Ischigualasto Formation, Argentina 32
isotope analysis 150-1, 154, 155, 206, 213

J

jaws 138-41, 143, 149, 158-9, 161, 212
Jerison, Harry 113
Jurassic 12, 14, 16, 18, 24, 34, 35, 36, 37, 38, 40-3, 46, 66, 87, 92, 95, 98, 103, 114, 135, 137, 154, 178, 208, 214
Jurassic Park (1993) 49, 50

K

Kansas 19
Kelvin, Lord 15
Khaan mckennai 115
kinematics, study of 90-1
Knight, Charles R. 47
Kosmoceratops 193
Koyna Lake, Western Ghats, India 204
Kulindadromeus 58, 59, 60, 172

L

LAGs 188, 212
Lambeosaurus lambei 195
Lark Quarry, Australia 82, 83
Late Cretaceous 16, 19, 84, 135, 137, 161, 206, 207
Late Jurassic 35, 66, 87, 92, 103, 116, 154
Late Permian 33-4
Late Triassic 15, 24, 32, 37, 38, 80, 83, 92, 208
legs, evolution of 36-7
lekking behavior 175
Leptorhynchos 134
Limaysaurus 121
Limusaurus 151, 154
lines of arrested growth 188, 212
lizards 22, 33, 52, 54, 57, 62, 68, 70, 76, 98, 124, 131, 145, 158, 165, 176, 198
Lockley, Martin 84, 175
locomotion
  flight 24, 40, 42, 43, 59, 98-103
  footprints/tracks 82-7
  gait 76-81, 90-1
  modeling 78-9, 212
  posture 47, 66, 76-81, 83, 92-3, 126-7
  speed 88-9

M

Mahakala 130
Maiasaura 168, 180, 181
Majungasaurus 140
Mamenchisaurus 95
mammals
  body temperatures 53, 54, 56
  brain 106, 110, 111, 112, 113
  breathing 64, 65
  coloring 172
  diet and size 134, 135, 137
  extinction 198, 208-9
  hearing 125
  posture 76, 77
  reproduction 176
  sexual dimorphism 167
  teeth 139, 140
Maniraptora 38, 43, 51, 115, 130
Maraapunisaurus 95
Marginocephalia 39
marine reptiles 19, 21-3, 33, 146, 198
Marsh, Othniel C. 118, 119
Martin, Tony 145
mass extinction 33, 34, 198-207
mass homeothermy 54
mating 166, 167, 174-5
MBD (multibody dynamics) 141
McNeil Alexander, Robert 88-9
Megalosaurus 46
melanin 172, 173
Mesozoic 12, 16, 21, 22, 24-5, 60, 152, 154
metabolic rate 53, 56, 57, 64, 65, 96-7
meteorites 198-9, 202, 203, 204, 206, 207
Mexico 19, 198-9, 202, 206
Mickelsen, Debra 87
Microraptor 40, 42, 43, 51, 60, 98, 100, 101-2
migration routes 84, 85, 86
miniaturization 40, 42-3
Monolophosaurus 151
Montana 19, 29, 66, 70, 71, 84, 86, 134, 136, 146, 181, 202
Morrison Formation, North America 66, 92, 154, 155
mosasaurs 17, 198
multibody dynamics (MBD) 141
Muttaburrasaurus 82, 83

221

N

Nannopterygius 21
Neoaves 115
nests/nesting grounds 179-81, 184
New Caledonian crow 107
New Mexico 15, 86
niche partitioning 153-5
Nicholls, Bob 51
nocturnal behavior 123, 125
North America 12, 16, 18, 19, 84, 92, 135, 203, 207
North Atlantic 16, 18, 34
North Dakota 202, 203

O

Oklahoma 19
olfactory bulbs 106, 110, 111, 120-1, 128, 129
Olorotitan arharensis 195
omnivores 134, 150, 151
origins, dinosaurs 32-5
ornithischians 35, 37, 39, 59, 60, 134, 139
ornithomimids 161
ornithopods 37, 39, 81, 83, 86, 114, 138
ostriches 53, 77, 88, 115, 123, 175, 185, 188
Oviraptor 178, 179
oviraptorosaurs 115, 130, 134, 135, 158, 161, 178

P

pachycephalosaurs 37, 39
Pachycephalosaurus 10, 70, 71
Pachyrhinosurus 192
pack hunting 161
paddlefish 202
Palaeeudyptes 208, 209
Palaeozoic 12
paleoartists 51, 172, 211
Paleogene 38, 206, 208-9
paleontology 10, 30-1, 46, 47, 214
Palmer, Colin 101
Pangea 16
parasagittal gait/posture 34, 37, 77
Parasaurolophus 191, 194, 195
parasites 60, 142, 147, 149

Paraves 115
paravian theropods 98
parental care 164, 176, 178, 180-3
peacocks 166-7
Pei, Rui 98, 99, 102
Pelecaniformes 115
Permian 12, 14, 16, 33, 34, 35
Phorusrhacos 209
physiology 46-67, 213
Piciformes 115
plantigrade stance 36
plants 152-3
plate tectonics 19
Plateosaurus 11, 92
Plesiadapis 209
plesiosaurs 21, 22, 25, 49, 198
pneumatization 64-6
posture 47, 66, 76-81, 83, 92-3, 126-7
Prasad, Vandana 146
Precambrian 12
primates 208, 209
Procellariiformes 115
Prosauropoda 38
Protoceratops 178
protorosaurs 145
Psittaciformes 115
Psittacosaurus 51, 60, 127, 174, 182, 183, 190, 211
pterosaurs
    extinction 198
    feathers 24-5, 35, 59, 60
    food webs 68
    hearing 131
    origins 33
    size 11, 24, 49, 213
    tracks 84

Q

Qi Zhao 182
Qin, Zichuan 161
quadrupedal posture 81, 83, 92, 126-7
Quetzalcoatlus 24, 202

R

radiometric dating 15, 32
Rahonavis 98, 100
Ratitae 115
Rayfield, Emily 141, 161
Regaliceratops 193
reproduction 164-71, 174-9
research, role of 210-13
respiratory diseases 66-7
rocks, order/age 12, 15
Romillo, Anthony 83
Rothschild, Bruce 147
running speed 88-9
Russell, Dale 116, 117

S

Saltasaurus 179, 184, 186, 187
Sander, Martin 96
Sapeornis 11
Saurischia 37, 38
Sauropoda 38
Sauropodomorpha 37, 38
sauropods
    breathing 64
    diet 134, 135, 154-5
    intelligence 114
    nests/nesting grounds 179, 184
    niche partitioning 154, 155
    origins 35
    posture 81
    size 92-7, 186
    smell (sense) 121
    teeth 138
    tracks 83, 84
scanning 62, 90, 108-9, 141, 172, 211
scavenging 159-60
Schaeffer, Joep 158-9
Schroeder, Kat 135, 136-7
scientific research 210-14
sea levels, changes in 16, 19
seed dispersal 152-3
seismic waves 203
senses 120-31
sexual dimorphism 167-8, 191
sexual selection 166-71
Shantungosaurus 11, 36
Shuvuuia 124, 125
sight 107, 110, 122-3
Similicaudipteryx 60
Sinosauropteryx 51, 58, 60, 61, 170, 171, 172

sinraptorids 136
size 10-11, 88-9, 92-7, 112-13, 134-7, 184-9
smell (sense) 110, 120-1
social behavior
    communicating 190-5
    eggs 176-9, 181, 184, 185
    herds 190-3
    hunting 159-61
    mating 164, 166, 167, 174-5
    parental care 180-3
    sexual selection 166-71
solitary hunting 161
South America 16, 19, 209
South Atlantic 16, 18, 19
spatial orientation 126-31
specialist diets 161
speciation rates 206
speed 7, 34, 77, 88-9
spinosaurids 160, 161
Spinosaurus 21, 43, 66, 81, 92
sprawling gait/posture 76
stampedes 82, 83
stegosaurs 37, 39, 81, 83, 114, 138, 168
Stegosaurus 10, 50, 66, 118, 119, 167, 168
Steneosaurus 214
Stenonychosaurus 116, 117, 123
stomach contents 71, 134, 144-5
stride length, and speed 88-9
Struthiomimus 77, 115
sturgeon 203
Styracosaurus 165, 192
Supersaurus 95

T

Tanis, North Dakota 202, 203
tapeworms 149
teeth 138-41, 143, 149, 158-9, 212
Tendaguru, Tanzania 92
Tenontosaurus 161
Tetanurae 38
Texas 16, 19, 24, 84, 86
Thecodontosaurus 108, 109
therizinosaurs 134, 161
Theropoda 37, 38
theropods
    birds 40-3, 98, 102
    brains 106, 107, 114
    breathing 66
    courtship 175

diet 134, 143, 151, 158-9, 161
feathers 59, 60
origins 34, 35
parental care 178
posture 37, 77, 78, 79, 81, 84
senses 121, 122-3, 125, 128
size 92, 135-7
teeth/jaws 141
tracks 83, 84
Thescelosaurus 202
Thyreophora 39
Titanoceratops 193
Torosaurus 11
tracks 82-91, 175, 191
Triassic 12, 14, 16, 18, 21, 32, 33, 34, 35, 36, 37, 38, 59, 77, 80, 208
Triceratops 7, 17, 19, 70, 71, 108, 114, 142, 153, 188
Trichomonas 147
Troodon 117
Troodontidae 115
Tsaagan 107
tsunamis 199, 202
Tupandactylus 25
turtles 22, 29, 57, 70, 84, 86, 124, 131, 168, 176, 185, 202
tyrannosaurids 136
Tyrannosaurus rex 7, 19, 28, 29, 43, 50, 66, 70, 71, 81, 88, 89, 92, 94, 114, 115, 123, 124, 128-9, 134, 135, 139, 141, 142, 143, 147, 149, 158, 159-60, 188, 189, 213

U

United States 40, 84, 85, 199, 202

V

Valley of the Moon, Argentina 32
Velociraptor 126, 130
Vinther, Jakob 172, 174
visual communication 191-3
visualizing dinosaurs 46-51
vocal communication 190, 194-5
volcanic eruptions 33, 34, 182, 183, 204, 206
vomit 145

W

walking 36, 37, 78-9, 82-7, 89, 90-1
Walking with Dinosaurs (1997) 49
warm-blooded animals 22, 23, 24, 34, 52-7, 97, 188
Waterhouse Hawkins, Benjamin 47, 49
Wegener, Alfred 16
weight, and flight 40, 43, 65
Weishampel, David 194, 195
Wellnhopterus 24
Western Interior Seaway 16, 17, 19, 84, 85, 86, 203
Wiemann, Jasmina 54, 177
Witmer, Larry 117, 128, 194
Wolff, Ewan 66, 147
Woodruff, Cary 66
Wyoming 16, 19, 87, 202

X

Xinjiangtitan 10
Xu Xing 60

Y

Yi qi 98, 101
Yinlong 151

Z

Zhao Qi 182
Zimmer, Carl 149

PICTURE CREDITS

Special thanks to: **62–3** Lida Xing, China University of Geosciences, Beijing, **84** Courtesy of Anthony R. Fiorillo, **90–1** Stephen Gatesy at Brown University, Rhode Island, **107** Comparing Dinosaur and Bird Brains: WitmerLab at Ohio University, **109** Infographic created by Antonio Ballell Mayoral. Original *Thecodontosaurus antiquus* silhouette by Jaime Headden, **125** Jonah Choiniere at University of the Witwatersrand, Johannesburg, **127** Mike Benton, **128** *Archaeopteryx* endocast by Amy Balanoff, **141** Emily Rayfield, University of Bristol, **148** Ewan Wolffe, **153** Seed photographs courtesy of Leonardo Salgado et al, **173** Klara Nordén, **183** Juvenile *Psittacosaurus* courtesy of Dr Qi Zhao (IVPP, Beijing, China), **187** Titanosaurian embryo photographs courtesy of Martin Kundrat, **202** Melanie During

13 Mx. Granger, CC0, via Wikimedia Commons, **15** slowmotiongli/Adobe Stock, **28** Puwadol Jaturawutthichai/Shutterstock, **31** Roderick Chen/All Canada Photos/Alamy Stock Photo, **41** James L. Amos, CC0, via Wikimedia Commons, **42** Lou-Foto/Alamy Stock Photo, **48–49** Chris Sampson, CC BY 2.0 creativecommons.org/licenses/by/2.0, via Wikimedia Commons, **50** top United Archives GmbH/Alamy Stock Photo, **50** bottom ScreenProd/Photononstop/Alamy Stock Photo, **53** Orhan Cam/Shutterstock, **61** Sam/Olai Ose/Skjaervoy from Zhangjiagang, China, CC BY-SA 2.0 creativecommons.org/licenses/by-sa/2.0, via Wikimedia Commons, **108** Dilip Vishwanat, **117** Dale A. Russell and Ron Séguin © Canadian Museum of Nature, **139** top stockdevil/Adobe Stock, **139** bottom Francois Gohier/ardea.com/agefotostock, **144** Tiouraren (Y.-C. Tsai), CC BY-SA 4.0 creativecommons.org/licenses/by-sa/4.0, via Wikimedia Commons, **145** Holgado B, Dalla Vecchia FM, Fortuny J, Bernardini F, Tuniz C (2015) A Reappraisal of the Purported Gastric Pellet with Pterosaurian Bones from the Upper Triassic of Italy. PLoS ONE 10(11): e0141275. https://doi.org/10.1371/journal.pone.0141275, **147** United States Geological Survey, Public domain, via Wikimedia Commons, **148** Wolff EDS, Salisbury SW, Horner JR, Varricchio DJ (2009) Common Avian Infection Plagued the Tyrant Dinosaurs. PLoS ONE 4(9): e7288. https://doi.org/10.1371/journal.pone.0007288, **173** 1. Brown-headed cowbird: ksblack99/Public Domain/Flickr, 2. Nicobar pigeon: Vassil, CC0, via Wikimedia Commons, 3. Elegant trogon: ALAN SCHMIERER/Public Domain/Flickr, 4. Variable sunbird: Leonard A. Floyd/Public Domain/Flickr, 5. Ruby-throated hummingbird: ksblack99/Public Domain/Flickr, 6. *Phasanius colchius*: Save nature and wildlife/Shutterstock, 7. *Anas fulvigula*: Sharon Wegner-Larsen/Public Domain/phylopic.org, 8. Trogon, 9. Hummingbird, and 10. Starling: Ferran Sayol/Public Domain/phylopic.org, **178** Dinoguy2, CC SA 1.0 <http://creativecommons.org/licenses/sa/1.0/>, via Wikimedia Commons, **205** Bert Willaert/Nature Picture Library/Alamy Stock Photo, **212** top JOSEPH NETTIS/SCIENCE PHOTO LIBRARY, **212** bottom Felix Choo/Alamy Stock Photo

ACKNOWLEDGMENTS

The author would like to thank the team at UniPress for their thorough and amazing work in bringing the book together. Thanks especially to Kate Duffy and Kate Shanahan for commissioning the book and for early planning, and to Katie Crous for her patience in editing and revising the text and images throughout the gestation of the book.

And, of course, thanks to Bob Nicholls, for bringing the book to life through his incredible artwork.

Bob Nicholls would like to thank Mike Benton for his continued support and confidence in his palaeo-artworks. Thanks to Katie Crous and Alex Coco for being so easy to work with and patient when the artworks were overdue. Bob also thanks Victoria, Darcey, and Holly for getting on with life while Daddy painted dinosaurs day and night and weekends for months on end. He owes all three of you a nice holiday and some chocolate!